Autonomous Safety Control of Flight Vehicles

T0132535

Autonomous Safety Control of Flight Vehicles

Xiang Yu
Lei Guo
Youmin Zhang
Jin Jiang

CRC Press
Taylor & Francis Group
Boca Raton London New York

CRC Press is an imprint of the
Taylor & Francis Group, an **informa** business

First edition published 2021
by CRC Press
6000 Broken Sound Parkway NW, Suite 300, Boca Raton, FL 33487-2742

and by CRC Press
2 Park Square, Milton Park, Abingdon, Oxon, OX14 4RN

© 2021 Xiang Yu, Lei Guo, Youmin Zhang and Jin Jiang

CRC Press is an imprint of Taylor & Francis Group, LLC

Library of Congress Cataloging-in-Publication Data

Names: Yu, Xiang, 1981- author.
Title: Autonomous safety control of flight vehicles / Xiang Yu, Lei Guo,
   Youmin Zhang, Jin Jiang.
Description: First edition. | Boca Raton, FL : CRC Press, an imprint of
   Taylor & Francis Group, 2021. | Includes bibliographical references. |
   Summary: "Aerospace vehicles are regarded as one classical type of
   safety-critical systems. By virtue of a safety control system, the
   aerospace vehicle can maintain high performance despite of component
   malfunctions and multiple disturbances, thereby enhancing the aircraft
   safety and mission success probability. This book presents a systematic
   methodology for improving the safety of aerospace vehicles in such
   aspects: the loss of control effectiveness of actuators and control
   surface impairments, disturbance observer-based control to against
   multiple disturbances, actuator faults and model uncertainties in
   hypersonic gliding vehicles, and faults composed by actuators faults and
   sensor faults. Several fundamental issues related to safety are
   explicitly analyzed according to aerospace engineering system
   characteristics. Focusing on the safety issues, the safety control
   design problems of aircraft are studied and elaborated in detail by
   using systematic design methods. The research results illustrate the
   superiority of the presented safety control approaches. The expected
   reader group for this book is undergraduate and graduate students but
   also industry practitioners and researchers"-- Provided by publisher.
Identifiers: LCCN 2020040179 (print) | LCCN 2020040180 (ebook) | ISBN
   9780367701154 (hardcover) | ISBN 9781003144922 (ebook)
Subjects: LCSH: Airplanes--Automatic control--Safety measures. | Flight control.
Classification: LCC TL589.4 .Y79 2021 (print) | LCC TL589.4 (ebook) | DDC 629.132/6--dc23
LC record available at https://lccn.loc.gov/2020040179
LC ebook record available at https://lccn.loc.gov/2020040180

ISBN: 978-0-367-70115-4 (hbk)
ISBN: 978-1-003-14492-2 (ebk)

To our families

Contents

Preface

Current trends toward greater complexity and automation are leaving modern technological systems increasingly vulnerable to faults and disturbances. Without proper action, even a minor error/disturbance may lead to destructive consequences. Over the past decades, the growing demand for safety, reliability, maintainability, and survivability in aerospace systems and industrial processes has motivated significant research in fault-tolerant control and anti-disturbance control. Note that fault-tolerant control and anti-disturbance control are named *safety control* in this monograph.

A control system that can accommodate faults among system components automatically, while ensuring system stability along with a satisfactory degree of overall performance, is named a *fault-tolerant control system* (FTCS). There are a broad range of applications of FTCSs in aircraft, space vehicles, power plants, and industrial plants processing hazardous materials, to name a few. In order to ensure the prescribed control performance, an anti-disturbance control system is designed with consideration of disturbance characteristics. Anti-disturbance control approaches can be further classified into active disturbance rejection control (ADRC), disturbance-observer-based control (DOBC), and composite hierarchical anti-disturbance control (CHADC), respectively.

Aerospace vehicles are regarded as one classical type of safety-critical systems. By virtue of a safety control system, the aerospace vehicle can maintain high performance despite of component malfunctions and multiple disturbances, thereby enhancing the aircraft safety and mission success probability. This monograph presents a systematic methodology for improving the safety of aerospace vehicles. Several fundamental issues related to safety are explicitly analyzed according to aerospace engineering system characteristics. Focusing on the safety issues, the safety control design problems of aircraft are studied and elaborated on in detail by using systematic design methods. The research results illustrate the superiority of the presented safety control approaches.

This monograph offers readers deep understanding and insights on system safety and the safety control design methods. The safety control methods proposed in this monograph can instruct engineers in improving the safety of aerospace engineering systems and promoting new technologies.

Structure and readership. This monograph introduces the design philosophies and methods of safety control systems with application to aerospace engineering systems. Chapter 1 introduces the development of safety control systems, including FTCS and anti-disturbance control systems. This chapter introduces the basic concepts of the FTCSs and compares the active FTCSs with the passive ones. Anti-disturbance control and the basic definition are also described.

The loss of control effectiveness of actuators and control surface impairments. In Chapter 2, in order to address the issue that the recovery time is limited after fault occurrence, a hybrid FTCS is proposed to accommodate the loss of control effectiveness of actuators. A design technique of an active FTCS against aircraft structural impairments is provided in Chapter 3. The control surface impairments are modeled as an LPV polytopic model using a parameter which is proportional to the effective control surface, and safety control laws are designed using both the state feedback and static output feedback against various degrees of control surface impairments.

Disturbance-observer-based control against multiple disturbances. Chapter 4 presents a multiple observers based anti-disturbance control (MOBADC) scheme against multiple disturbances for a quadrotor unmanned aerial vehicle (UAV). The quadrotor UAV dynamics can be represented by Newton's second law and the Lagrange-Euler formalism. The proposed control scheme consists of a disturbance-observer-based controller and an extended state observer (ESO) based controller, which are utilized in a position loop to mainly eliminate the cable-suspended-payload disturbance with partially known information and mitigate the wind disturbance with bounded variation. Furthermore, in order to reject the model uncertainty and disturbance moment, another ESO-based controller is designed for the attitude loop.

Actuator faults and model uncertainties in hypersonic gliding vehicles. Chapter 5 integrates the multivariable integral terminal sliding mode control (TSMC) and adaptive techniques to handle the actuator malfunctions and model uncertainties for a hypersonic gliding vehicle (HGV). In Chapter 6, in order to counteract actuator faults and model uncertainties, a fixed-time observer is designed to ensure the disturbance observer is independent of initial conditions and the estimation errors for converging to zero in fixed settling time. Subsequently, the finite-time multivariable TSMC and composite-loop design are pursued to guarantee the safety of the vehicle in a timely manner.

Faults composed by actuators: faults and sensor faults. In Chapter 7, a fault accommodation scheme with consideration of actuator control authority and gyro availability is developed with the integrated design of a sliding mode observer (SMO) and sliding mode control. Fault accommodation can be promptly accomplished and decouple difficulties can be addressed.

Acknowledgements. The authors would like to express their sincere appreciation for colleagues' valuable suggestions and help. This book was supported by the National Natural Science Foundation of China (Project numbers 61833013 and 61973012), the Program for Changjiang Scholars and Innovative Research Team (Project number IRT 16R03), Fund of Key Laboratory of Defense Science and Technology for Integrated Aircraft Control Technology WDZC2019601A104, Zhejiang Lab (No. 2019NB0AB08), and NSERC.

Beihang University, Beijing, China   Xiang Yu

Beihang University, Beijing, China   Lei Guo

Concordia University, Montréal, Canada   Youmin Zhang

Western University, London, Canada   Jin Jiang

List of Figures

List of Tables

Chapter 1

The Development of Safety Control Systems

1.1 Introduction

Over the last few years, the growing demand for safety, survivability, and high-precision control performance in aerospace engineering systems has motivated significant research on fault-tolerant control systems (FTCSs) and anti-disturbance control systems (ADCSs). The major objective of FTCSs and ADCSs, named as safety control systems (SCSs) in this monograph, is to effectively handle abnormal situations under more realistic aerospace vehicle operations where both faults and disturbances could occur simultaneously, so that the safety can be guaranteed. An autonomous SCS should consist of disturbance/fault analysis, task reconfiguration, and safety control law, respectively. Before designing an autonomous SCS, it is of paramount importance to analyze the impact of uncertainty, disturbances, and faults that may induce catastrophic consequences. The system may no longer accomplish the preassigned task, due to the capability of the overall system is degraded. In this case, task should be reconfigured according to the quantitative analysis of capability. Subsequently, safety control law is triggered autonomously to ensure the safety of the system. Note that this monograph focuses on the disturbance/fault analysis and safety control law design. In the following, the existing FTCS and ADCS approaches are reviewed.

A control system that can accommodate faults among system components automatically while maintaining system stability along with a desired level of overall performance is denoted as a FTCS [1, 2, 3, 4]. Several approaches of designing FTCSs have been developed for safety-critical applications [1, 2, 3, 4].

In an FTCS, the achievable system performance depends on the availability of redundancies in the control system as well as the design approaches used in the synthesis of fault-tolerant controllers. Depending on how redundancies are being utilized, current FTCSs can be classified into two categories, namely, active FTCSs [5, 6, 7, 8, 9, 10, 11, 12, 13, 14, 15, 16, 17], and passive FTCSs [18, 19, 20, 21, 22, 23, 24, 25, 26, 27]. These two approaches use different design methodologies for the same control objective. Even though, as far as the main control objectives are concerned, both methods lead to similar results, however, due to the distinctive design approaches used, each method can result in some unique properties.

A systematic development of fault-tolerant control (FTC) is presented in [1]. The authors consider the entire design process for FTC from engineering of interfaces to structural implementation. A temperature control loop for a fluid cooling system and an attitude control system for a satellite are selected as examples to illustrate the development process. The state-of-the-art of FTCS technologies for aerospace systems has been examined [3]. It provides a comprehensive literature review covering most areas of FTC. Design of FTC, role of FDD unit, and the interaction between the FTC and the FDD are investigated. Based on the concept of redundancies, an introductory overview on the development of FTCS is presented by [2] from a practical and industrial perspective. The analysis techniques for active and passive FTCSs are listed. By using several practical applications, such as commercial jets and nuclear power plants, the relationships among redundancy, safety, and performance are explained. The philosophies of active and passive FTCSs are given as well as the potential challenges. [4] presents an extensive bibliographical review on the historical and current development in active FTCS. The motivation, objective, and structure of active FTCS are discussed. The existing methodologies on FDD and reconfigurable control are classified based on algorithm and field of applications. The features of current techniques are briefly summarized.

In addition to malfunctions, multiple disturbances widely exist in any aerospace vehicles. Multiple disturbances create great difficulties of achieving high-precision control performance and high degree of safety [28]. The disturbances and noises can be characterized as an uncertain norm-bounded variable, a harmonic, a step signal, a non-Gaussian/Gaussian random variable, a variable with bounded change rate, output variables of a neutral stable system, and other types of disturbances. Furthermore, multiple disturbances arising from multiple resources are exposed on various channels of aerospace engineering system. Depending on the different places where disturbances act on, disturbances can be further categorized into internal disturbances, external disturbances, and model uncertainties, respectively. In an effort to enhance both the performance and safety, the issue of disturbance attenuation or rejection has drawn tremendous research interest in aerospace engineering community.

1.2 Philosophical Distinctions between Active and Passive FTCSs

The main objectives of an FTCS are to preserve the stability of the overall system and to maintain an acceptable level of performance in the event of system component malfunctions. Component failures can be divided into two types: those that can be anticipated at the design stage and those that are out of pre-defined fault sets, and occur only during system operations. The approaches to deal with failures can also be different. One is to respond to the failure by re-organizing the remaining (often redundant) system elements in real-time to carry out necessary control functions. The other is to make the system failure-proof for a certain well defined fault sets at the design stage by using built-in system redundancies. The first approach leads to an active FTCS and the second to a passive one. A commonality in both methods is the existence of system redundancies, and difference is how the redundancy is utilized.

1.2.1 Architecture and Philosophy of an Active FTCS

An active FTCS reacts to system component malfunctions (including actuators, system itself, and sensors) by reconfiguring the controller based on the real-time information from an FDD scheme. The term 'active' represents corrective actions taken actively by the reconfiguration mechanism to adapt the control system in response to the detected system faults. As shown in Fig 1.1, an active FTCS typically consists of an FDD scheme, a reconfigurable controller, and a controller reconfiguration mechanism. These three units have to work in harmony to complete successful control tasks. Based on this architecture, the design objectives of an active FTCS are (1) to develop an effective FDD scheme to provide information about the fault with minimal uncertainties in a timely manner; (2) to reconfigure the existing control scheme effectively to achieve stability and acceptable closed-loop system performance; and (3) to commission the reconfigured controller smoothly into the system by minimizing potential switching transients.

The philosophies of an active FTCS can be demonstrated in Fig 1.2, where both the fault-free and fault cases are considered. For each case, it is assumed that there is a corresponding admissible solution space, in which a controller exists. Within each of these admissible regions, an optimal solution with certain pre-set performance criteria can be found by FTC synthesis methods. The existence of these optimal solutions depends on the constraints and optimization techniques used. Once a solution is found, the new controller can be put into action through the reconfiguration process to compensate the effects of the fault. The controller design algorithm used in an active FTCS can be made to search for an optimal solution under specific fault considerations.

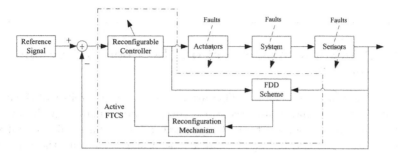

FIGURE 1.1: Architecture of an active FTCS.

However, in practice, a sub-optimal controller would be sufficient, because it will take more time to obtain the optimal solution. In the presence of a fault, the amount of time available to implement a corrective control action is very limited. As can be seen from Fig 1.2, different controllers in an active FTCS can be designed within the admissible sets in response to specific faults. Hence, an active FTCS has the potential to achieve best possible performance if the diagnostic information from the FDD scheme is accurate and timely, and there is sufficient amount of reaction time for controller synthesis.

Even though the above concept sounds straightforward, there are other issues which can impair an active FTCS to achieve its mission. The transitions among the admissible sets shown in Fig 1.2 depend on several factors, namely, the time needed to obtain the fault information, the accuracy of the fault information, and the time taken to synthesize and deploy the new control schemes. It should be pointed out that the time taken by an FDD process is highly dependent on the nature of faults, prescribed uncertainty bounds, and selected FDD algorithms [29]. Often, system component faults can render the system unstable, and therefore there is only a limited amount of time

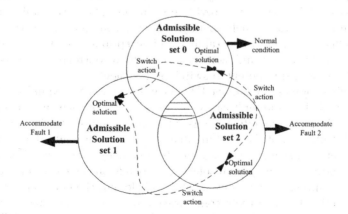

FIGURE 1.2: Admissible solution space for active and passive FTCSs.

available to react to the faults. From an operational point of view, an active FTCS represents a safety-critical real-time system. After a fault occurs, there is normally a limited amount of time available for an active FTCS to react to the fault and to make corrective control actions. This time can be referred to as 'critical reaction time' (T_c). If the true reaction time of the active FTCS is longer than the critical reaction time, the system enters an unrecoverable state. These two situations are shown in Fig 1.3(a) and Fig 1.3(b), respectively. The total reaction time of an active FTCS (T_a) consists of the time needed for FDD (T_{FDD}) and the time for the controller reconfiguration (T_{RC}). Similar 'time-critical' situations in safety-critical system applications can be found [5, 7, 13, 15, 16], where an FTCS is warranted.

1.2.2 Architecture and Philosophy of a Passive FTCS

In a passive approach, a list of potential malfunctions is assumed known a priori as design basis faults, and all failure modes as well as the normal system operating conditions are considered at the design stage. Neither an FDD scheme nor a controller reconfiguration mechanism is needed. Therefore, the term 'passive' indicates that no additional actions need to be taken by the existing control system in response to the design basis faults. The controller

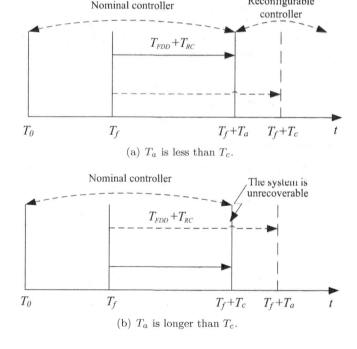

(a) T_a is less than T_c.

(b) T_a is longer than T_c.

FIGURE 1.3: Critical reaction time of an active FTCS.

deals with the faults passively. As shown in Fig 1.4, a passive FTCS is a control system designed to tolerate system component faults by using the system redundancies without any controller structure or parameter adjustment. The objective of a passive FTCS design is to synthesize a single fixed controller to make the closed-loop system as insensitive as possible to the set of design basis faults.

The philosophy of a passive FTCS is to find a controller within the region of intersection among all admissible solution sets. As shown in Fig 1.2, this region corresponds to the shadowed area where the admissible solution sets intersect. Several limiting situations in a passive FTCS synthesis are shown in four sub-plots in Fig 1.5 where no single overlap can be found. Sacrifices may have to be made even for the normal system operating conditions to accommodate anticipated failure cases. As shown in Fig 1.5(c), no passive FTCS can be found to deal with the normal condition and the pre-considered faults 1 and 2 simultaneously.

From performance perspective, a passive FTCS focuses more on the robustness of the control system to accommodate multiple system faults without striving for optimal performance for any specific fault condition. In comparison with an active FTCS, it is more difficult for a passive FTCS to achieve optimal performance under any design basis fault condition. Since the stability is the number one consideration in a passive approach, the designed controller turns to be more conservative from performance viewpoint.

If there is an overlap among all the admissible solution sets for considered fault cases, a single controller can theoretically deal with any presumed design basis faults. However, nothing can be said about the behavior of the system when an failures beyond design basis occur. It is critical to emphasize that the number of the design basis faults that the FTCS can deal with also depends on the availability of the redundancies. Nevertheless, since a passive FTCS does not involve controller reconfiguration, there are no switching transients.

1.2.3   Summary of FTCS

The objectives of active and passive FTCSs against actuator failures are to compensate for the loss of control actions by appropriately reassigning the control signals to the remaining healthy control surfaces. The philosophies and architectures for both active and passive FTCSs are presented. Comparisons

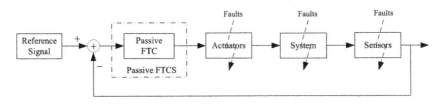

FIGURE 1.4: Architecture of a passive FTCS.

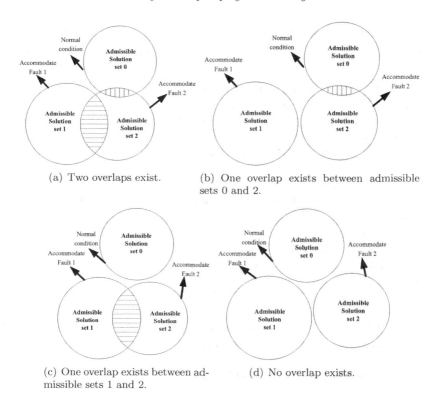

(a) Two overlaps exist.

(b) One overlap exists between admissible sets 0 and 2.

(c) One overlap exists between admissible sets 1 and 2.

(d) No overlap exists.

FIGURE 1.5: Illustration of overlap among admissible solution sets.

between the two approaches are made in the following. The characteristics of both active and passive FTCSs are then summarized. Through analysis and comparison of the simulation results, the essential characteristics of two FTCS strategies are summarized in Table 1.1. The detailed information is given as follows.

1.2.3.1 Advantages of an Active FTCS

Active FTCS can reconfigure the controller by using the real-time information provided by the FDD scheme. From the conceptual illustrations in Fig 1.2, an active FTCS initially operates in the normal mode using an admissible solution with certain performance levels. Upon detection of a fault, it moves to another admissible solution region to counteract the fault effects. Within each of these admissible regions, the optimal solution may exist with certain pre-set performance criteria.

Another advantage of an active FTCS is that it can deal with beyond design basis failures as long as the FDD scheme can detect and diagnose them correctly in time, and also there exists a sufficient degree of redundancy to

TABLE 1.1: Comparison of the characteristics of an active and a passive FTCSs.

	Active FTCS	Passive FTCS
Potential for performance optimization	Yes	No
Dealing with beyond design basis failures	Yes	No
Immediate control action after the fault	No	Yes
Sensitive to the results of FDD	Yes	No
Guaranteed stability for the design basis faults	N/A	Yes
Switching transients	Yes	No
Smooth in operation during a fault occurrence	No	Yes
Time before control in action	Yes	No
Easy in implementation	No	Yes
Controller design time (based on optimization)	Short	Long

make up the shortfall of the failed actuators. However, the accuracy and the time taken for the diagnosis are the key.

1.2.3.2   Limitations of an Active FTCS

In the current active FTCS design, no action is performed until the new control system is synthesized. However, the interval between the fault occurrence and the initiation of the reconfigured controller plays a role in maintaining the safe operation of the system. If the FDD process takes an unexpectedly long time, the integrity of the system may be in danger. It is very likely that the performance of the system deteriorates with an increase in FDD time. In some safety-critical systems, certain critical system variables may cross the safety boundaries if no actions are taken within a certain critical period of time.

The reconfigurable controller design relies heavily on the information of the fault provided by the FDD scheme. The performance of an active FTCS is highly dependent on the accuracy of FDD. Any uncertainties in FDD can lead to loss of effectiveness in the designed controller. And an error in FDD can have a significant impact on the quality of the active FTCS results. The altitude of the aircraft cannot even be maintained in the event of these uncertainties. In addition, the overall system performance can degrade progressively as the magnitude of the uncertainties increases. In an active FTCS, a newly reconfigured controller has to be switched in to replace the pre-fault controller. The switching transients, which are essentially shocks to the system, are highly undesirable and can potentially lead to further damage to the system components.

1.2.3.3 Advantages of a Passive FTCS

In a passive FTCS the controller, once designed, does not need to be changed during the course of operation. In practice, a passive FTCS has a simple structure and has no controller switching associated transients. Therefore, the additional real-time computational demand is low for a passive FTCS.

Since no switching is involved in a passive FTCS, the behavior of the system is much smoother than that of an active FTCS. Furthermore, since the passive FTCS does not require any FDD unit, there is no delay between the fault occurrence and the corresponding control actions. The control system is fully engaged, and the control actions to the fault occurrence are always immediate.

1.2.3.4 Limitations of a Passive FTCS

A passive FTCS is designed with the consideration of both normal system operation and design basis faults. Compared with an active approach, the performance achieved by a passive FTCS can never be optimal for all design scenarios. If one attempts to design a passive FTCS to accommodate excessive number of faults, the overall conservatism increases. No controller may be found to satisfy all the design requirements.

Since the philosophy of a passive FTCS is to find a region of intersection among several admissible solution sets, when the number of fault scenarios increases beyond a certain number, such a region of intersection may not even exist.

Compared with an active FTCS, a passive FTCS is less flexible and has limited fault-tolerant capabilities, especially in the case of beyond design basis failures. The overall performance of the controller becomes less and less effective for each fault case as the number of fault cases increases. The passive FTCS designed on the basis of cases can no longer guarantee the stability and acceptable performance when failures beyond the passive FTCS design basis has occurred.

The comparison of both active and passive FTCS strategies leads one to think whether it is possible to design a "hybrid" FTCS to combine the merits of active and passive FTCSs, and to discredit their respective disadvantages. In fact, such hybrid FTCS concept has been proposed [9, 30] to deal with actuator faults. In this concept, a passive FTCS is used to slow down the deterioration of the system with minimal amount of fault information. As more detailed fault information becomes available, effective reconfigurable controllers can be designed and subsequently switched to achieve improved system performance.

1.3   Basic Concept and Classification of Anti-Disturbance Control Systems

The control approaches against disturbances can be mainly divided into disturbance attenuation methods (such as robust control theory) and disturbance rejection schemes. The existing works can be generally classified into five types, which are robust control [31], adaptive control [32], active disturbance rejection control (ADRC) [33, 34], disturbance-observer-based control (DOBC) [35, 36, 37], and composite hierarchical anti-disturbance control (CHADC) [28, 38], respectively.

Within a robust control scheme, the detrimental effect resulting from disturbances can be handled in a "passive" manner. Adaptive control techniques are used in flight control design by accounting for model uncertainties and disturbances. The basic idea of adaptive control approaches is to adjust control gains according to aircraft dynamic variations induced by disturbances. The design philosophy of ADRC is to recast the "overall disturbance" as one derivative bounded variable, which is of interest for an extended state observer (ESO). ADRC plays an important role in rejecting disturbances with uncertain dynamics. In most cases, errors or disturbances are time-varying unknown dynamics, for which observers or filters can be developed for estimation. The primary concept of DOBC is to estimate the external disturbance by virtue of disturbance observer, and subsequently compensate in the feedforward channel. DOBC approaches have been widely applied in practical engineering systems. At first, linear DOBC (LDOBC) methods are utilized for linear single-input single-output (SISO) systems in the frequency domain. Furthermore, nonlinear DOBC approaches are proposed to improve the control accuracy. One feature of DOBC is with simple structure. Along with the development of sensor technology and data processing, one can formulate them into different mathematical descriptions after explicit analysis. CHADC can be regarded as a viable option of simultaneously attenuating and rejecting disturbances. The CHADC scheme usually consists of the disturbance observer(s) and the baseline controller.

Robust control or adaptive control has the following limits: 1) strict limitation of plant model; 2) complexity to hinder the practical implementation; and 3) lack of disturbance knowledge and analysis. ADRC theory has better prospects for engineering applications, but still lacks theoretical proof of the stability. Moreover, the treatment that multiple disturbances are considered as a lumped disturbance inevitably induces the conservativeness. Despite of the significant progress in DOBC, multiple disturbances are handled as a lumped disturbance. Nevertheless, the disturbance analysis and representation are inadequate.

1.4 Safety-Critical Issues of Aerospace Vehicles

Even though great efforts have been devoted in both FTC and anti-disturbance control communities to improve performance and safety, there still exist several safety-critical issues of aerospace vehicles.

1.4.1 Safety Bounds

Aerospace vehicle as a physical system has a specific capability (e.g., flight envelope). In particular, the states and actuator outputs of flight vehicles are restricted within allowable ranges, which can be also named safety bounds. As long as the safety bounds are breached, safety is shrunk and even worse crash is induced. There are two important aspects to system safety: 1) safety bounds of the aircraft variables need to be respected in the safety (fault-tolerant, anti-disturbance) control design, since the purpose of safety control is to maintain safety; 2) stringent safety bounds require that the safety control is robust against imperfect FDD or disturbance estimation information [39, 40]. In some urgent situations, it is required that maintaining aircraft safety can still be achieved without waiting for the final convergence of the FDD or disturbance estimation algorithm.

1.4.2 Limited Recovery Time

Safety-critical systems have a limited amount of time available for recovery when they are suffering malfunctions in operation [30, 41, 42, 43]. It should be emphasized that the time available for recovery (if recovery is possible) depends on the system state, the fault, and the time of fault occurrence [41]. On the other hand, the time needed by the FDD is closely tied with operating points, fault type, and the degree of severity of faults. In [30, 41, 42], the authors argue that each FDD module needs a specific amount of time to accomplish FDD tasks no matter what techniques are used. Moreover, control laws need time to be reconfigured, and as a result, the system is prone to abnormalities before the onset of the control reconfiguration. In Fig. 1.6(a), the faulty system cannot be rescued since the reconfiguration is commissioned too late and the system has already entered an irreversible state (i.e. the safety limits are violated). As illustrated in Fig. 1.6(b), if the available time for recovery is longer than that needed for active FTC then it is possible to ensure the safety of the system under failures.

1.4.3 Finite-Time Stabilization/Tracking

The finite-time control theory has been widely developed to enable system states to approach equilibrium within finite time. Focusing on aircraft safety,

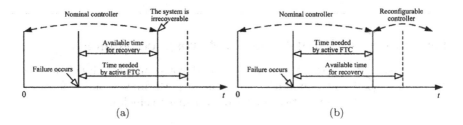

FIGURE 1.6: The safety issue relevant to time.

it is highly expected to ensure the safety in the presence of faults/disturbances as soon as possible, since the amount of time available for recovery is strictly limited in aerospace engineering practice. In the references [44], finite-time control method is applied for FTC design to guarantee the stability or tracking performance under actuator faults. In this research area, the recoverability [41, 45] needs to be taken into account, so that the physical limits of the faulty system are not violated.

1.4.4   Transient Management

The importance of transient management has been mentioned in [4, 41]. In an active FTC scheme, unfavourable transients may occur during the control reconfiguration process. The abrupt transients may result in a secondary damage of the healthy components, and worse still the break-down of the system. In the reference [46], hybrid system techniques are used to specify the transition management strategies. More recently, Lyapunov global exponential stability of switched systems [47] and an adaptive variable structure control [48] are exploited to improve the transient behaviors during fault accommodation. In this topic, how to integrate safety constraints into transient management should be explicitly investigated.

1.4.5   Composite Faults and Disturbances

In real aerospace applications, exhaustion of physical control authority is known as magnitude and/or rate limiting of the actuators. Actuator amplitude saturation is dangerous, as it does not only affect aircraft performance but also flight safety. Additionally, actuation rate bounds on a flight actuator are inevitable owing to the maximum flow capability of the main control valve and the flight-induced aerodynamic load. The literature concerning this topic can be dominated by two concepts: 1) the magnitude limits of redundant actuators are not violated by properly managing command/reference signals [11, 49, 50, 51]; 2) the control laws are reconfigured with the integration of actuator displacement bounds [52, 53, 54, 55, 56].

Sensors are identified as the weak link in aerospace engineering systems based on being more vulnerable to damage or being more sensitive in construction than other components [57]. And the failure rate of gyros is ranked the highest among the key components in an inertial measurement unit (IMU) [58]. To meet the stringent demands of aircraft reliability, physical redundancy in the form of identical sensors or multiple types of sensors is often configured [58]. Nonetheless, redundant sensors are not always feasible due to the constraints of weight, cost, space, power, and complexity. Thereby, analytical redundancy constituted by the knowledge of the flight vehicle becomes a viable supplement [59]. Therefore, the design of safety control systems, which can accommodate actuator malfunctions and ensure actuation amplitude and rates remain within the allowable bounds and handle sensor faults simultaneously, is still a challenge [60].

In addition to the composite faults of actuator and sensor, it is extremely difficult to diagnose and counteract the faults if disturbances and faults exist simultaneously in aerospace vehicles. The presence of disturbances has an inverse impact on FDD and subsequently on control reconfiguration. To date, it is difficult to isolate disturbance and fault. The coupling effects of both disturbance and fault should be systematically analyzed. Furthermore, more emphasis on designing safety control system in the presence of composite disturbance and faults must be placed.

1.5   Book Outline

In the monograph, from the perspective of practical engineering, we study the theory of FTCS and apply the FTC theory to the aircraft. The monograph is organized as follows. Chapter 1 reviews the development of safety control system. Active and passive FTCSs are introduced by examining the similarities and differences between these two approaches. Advantages and limitations of each method are examined through a philosophical level. And a hybrid FTCS that combines the merits of passive and active FTCSs is proposed to accommodate the loss of control effectiveness of actuators in Chapter 2. Chapter 3 provides a design technique of an active FTCS against control surface impairments, which considers both state feedback and static output feedback. In Chapter 4, a MOBADC scheme against for a quadrotor UAV is proposed to cope with multiple disturbances. Chapter 5 presents a safety control system for a HGV subject to actuator malfunctions and model uncertainties, based on adaptive multivariable TSMC technique. Chapter 6 proposes a design of fixed-time observer based safety control system of HGV against faults and uncertainties. In Chapter 7, a fault accommodation scheme with consideration of actuator control authority and gyro availability is developed.

Chapter 2

Hybrid Fault-Tolerant Control System Design against Actuator Failures

2.1 Introduction

Essentially, the safety of the system in the presence of component failures is closely associated with the availability of physical redundancies and fault-tolerant control strategies. Depending on how redundancies are utilized, existing FTCSs can be categorized into two types: passive and active strategies [2, 4, 61].

In a passive FTCS, all potential failure modes are assumed to be known a priori, and are considered together with the nominal operating conditions in the design process. Since no fault detection and diagnosis (FDD) unit or reconfiguration mechanism is required, a passive FTCS is simple to implement in practice. However, a major drawback is that the controller design becomes more and more conservative as the number of considered fault scenarios increases. An active FTCS reacts to specific system malfunctions by reconfiguring the controller based on the information from an FDD scheme, so that the stability and acceptable performance of the overall system can be maintained [62]. An active FTCS reacts to system component malfunctions more effectively than a passive FTCS does if the fault can be diagnosed correctly. The critical issue in any active FTCS is that there is only a limited time available to perform the FDD tasks. In practice, the fault can progress rapidly and the critical system states can even cross the system safety boundaries before a desirable active FTCS can be synthesized and commissioned. Therefore, it would be highly desirable if one can develop an FTCS to maintain the system stability with minimal fault information before complete fault information is available. Once the correct fault diagnostic information becomes available, it can then be used in controller design process for performance improvement.

In most of the existing active FTCS methods, the stability of the post-fault system cannot be guaranteed during the period that the FDD unit is performing its diagnostic functions. This problem has been investigated in Ref. [9]. In Refs. [63, 64, 65, 66], a fast switch from a normal operation mode to a safe or reconfigured operation mode can be realized by selecting an FTCS architecture, which is based on the Youla-Jabr-Bongiorno-Kucera (YJBK) parameterization. It allows a very fast switch from a normal mode controller to a safe mode controller by removing the reconfiguration loop in the controller architecture. In the current approach, switching from a nominal controller to a passive FTCS, and then to a reconfigurable controller is realized based on the information from FDD scheme and system performance requirements. The focus has been on switching triggered by the convergence of the estimation of the actuator effectiveness. The passive FTCS can guarantee the stability of the post-fault system and the active FTCS with the correct fault information can improve the performance. Both the reconfigurable controller and the passive FTCS are synthesized by using an LMI approach.

Actuators play an important role of linking control commands to physical actions performed on the system to achieve specific objectives. Normally, the actuators should execute commands demanded by the controller faithfully and completely [67]. Under this condition, the actuators are referred as 100% effective. However, when a fault occurs in the actuator, the handicapped actuator might not be able to complete the control command fully. Naturally, the control channel effectiveness (or lack of it) becomes an appropriate measure of the severity of the actuator fault. In an aircraft, a significant number of actuators are in the form of hydraulic driven control surfaces. Faults can cause discrepancies between the desired and the actual movements of these control surfaces. Potential root causes for such faults can be due to incorrect supply pressure in the hydraulic lines, change in hydraulic compliance, and line leakage [68, 69, 70, 71]. Any of these problems can prevent the primary control surfaces such as elevators, ailerons, or rudder from moving to the positions demanded by the controller.

The remainder of this chapter is organized as follows: modeling of actuator faults through control channel effectiveness is introduced with some discussions on actuator redundancy in Chapter 2.2. The hybrid FTCS problem formulation is addressed in Chapter 2.3 together with the description of the FTCS scheme and its related components. In Chapter 2.4, the design of a passive FTCS and a reconfigurable controller is elaborated. Simulations of the system based on linearized and fully nonlinear models are carried out and the results are presented in Chapter 2.5 to validate the proposed scheme, followed by some concluding remarks in Chapter 2.6.

2.2 Modeling of Actuator Faults through Control Effectiveness

2.2.1 Function of Actuators in an Aircraft

Actuators that link control commands to physical actions on an aircraft are essential elements in any flight control system. In an aircraft, hydraulic mechanisms are widely used for actuation purposes due to their high force to inertia ratio. For instance in an F-16 fighter, an F-18 fighter, a JAS-39 Gripen aircraft, and a Boeing 737, the actuators such as ailerons and elevators are all in the form of hydraulic driven control surfaces. The function of a control surface is to produce the required torque and moment to maneuver the aircraft. Different maneuvering commands are realized through deflecting appropriate control surfaces in various parts of the aircraft. When all components function normally, the desired control actions can be carried out exactly as the controller has demanded. However, when some failures occur in the actuators, the desired control commands cannot be completed as expected. The stability and the performance of the aircraft can suffer as a consequence.

2.2.2 Analysis of Faults in Hydraulic Driven Control Surfaces

A hydraulic driven actuator consists of three main parts: a hydraulic power supply, a servo-valve, and control surface. The power supply delivers hydraulic fluid to the high-pressure port of the servo-valve at a constant pressure. As shown in Fig. 2.1, the torque motor converts the corresponding voltage from the controller into the angular displacement on the baffle, which generates a differential pressure in the servo-valve. Subsequently, the servo-valve regulates the motion of the actuator by directing the fluid flow to and from the actuator chamber. When the valve spool is pushed or pulled by the torque motor that is driven by the control signal, the amount of fluid delivered to the actuator chamber will change. The fluid will then move the piston in the actuator chamber. Since the control surface is physically connected to the piston through the piston rod, the displacement of the piston leads to movement in the control surface. The torque generated by the force through a hinge to deflect the control surface also has to overcome the aerodynamic force during a flight.

The nonlinear relationship that describes the motion, D_{spool} of the valve spool in response to the displacement of the piston pod, D_{pod}, as well as the corresponding linearization procedures are elaborated in Appendix A. The physical meanings of the symbols used in the derivations are summarized in Table 2.1. In the current work, modeling uncertainties of the actuator have not been explicitly considered.

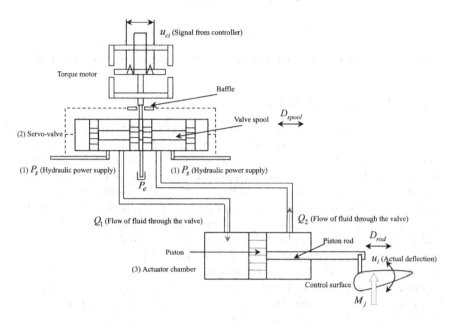

FIGURE 2.1: Schematic diagram of a hydraulic driven control surface.

A common failure in a hydraulic powered actuator is loss of pressure in the supply pump due to leaks [70]. Reduction in pressure could cause stalling of the actuator when it can no longer balance the aerodynamic load imposed from the control surface during a flight. In an aircraft whose control surfaces are manipulated through hydraulic actuators, stalling of an actuator may have disastrous consequences, because control actions would not get executed as expected. In this chapter, an incorrect supply pressure is considered as the fault condition to illustrate the hybrid FTCS design concept and process. Other failure scenarios can be treated similarly, but this is beyond the scope of the current chapter.

From Appendix A, the relationship between the control command u_{ci} and the actual deflection of the control surface u_i can be represented as

$$\frac{u_i(s)}{u_{ci}(s)} = \frac{k_{ha}k_v}{(T_{ha}s + 1)\left(\frac{s^2}{\omega_v^2} + 2\frac{\xi_v}{\omega_v}s + 1\right)}, \tag{2.1}$$

where

$$k_{ha} = \frac{k_1 FL}{M_j^\delta (C_1 + C_2)}, T_{ha} = \frac{k_M M_j^\delta \left(k_e + \frac{V_0}{E}\right) - \frac{2}{57.3}k_1 FL^2}{k_M M_j^\delta (C_1 + C_2)}.$$

Since the gain k_1 is proportional to the gain k_{ha} and the reduction of supply pressure can lead to the change of k_1, k_1 is used to analyze the control

TABLE 2.1: Nomenclature of hydraulic actuator.

Symbol	Physical Meaning
D_{spool}	Valve spool displacement
D_{rod}	Rod displacement
C_1	Pressure sensitivity gain of the valve
C_2	Effective discharge coefficient of the leakage orifice
F	Piston effective area
L	Length of rocker
M_j^{δ}	Hinge moment coefficient
M_j	Hinge moment (load of control surface)
V_0	Actuator cylinder column
E	Elastic modulus of oil
k_l	Flow sensitivity gains of the valve
k_{δ}	Control surface deflection gain
k_e	Pipe coefficient of elasticity
f	Piston damping coefficient
k_v	Valve spool position gain
τ	Time constant of the transfer function of the controller output voltage and the valve spool
u_i	Actual deflection
u_{ci}	Controller command
P	Load pressure
P_s	Supply pressure
ρ	Density of the hydraulic fluid
C_v	Valve coefficient of discharge
w	Valve orifice area gradient

effectiveness loss. From Ref. [72], the flow sensitivity gain of the valve is defined as:

$$k_1 = C_v w \sqrt{\frac{P_s - P}{\rho}}, \tag{2.2}$$

where C_v, w, P_s, P, and ρ represent the valve coefficient of discharge, valve orifice area gradient, supply pressure, load pressure at an operating condition, and the density of the hydraulic fluid, respectively. The nominal values of these parameters are provided in Ref. [72] as 0.6, 20.75 mm, 17.2 MPa, 11.5 MPa, and 847 kg/m^3. Based on Eq. (2.2), the nominal value of k_1 can be determined as 1.02.

When the supply pressure drops from nominal value, it means that a fault has occurred. As observed in Eq. (2.2), reduction in P_s induces loss in the steady-state gain of the actuator channel. For example, assuming the supply pressure is reduced from 17.2 MPa to 12 MPa, the post-fault value of k_1 will be reduced from 1.02 to 0.3025 accordingly. Since the steady-state gain $k_{ha}k_v$ of the actuator is proportional to k_1, this particular fault can be modeled as 29.66% loss in the effectiveness of the actuator.

It should be mentioned that when the actuator and aircraft are considered together as an integrated unit, the time constant of the actuator is much smaller than that of the aircraft [73]. Hence, the dynamics of the actuator can be ignored in the control system design process without causing significant error. As a result, only the gain of the transfer function of the actuator $k_{ha}k_v$ is used to characterize the effectiveness of the control action. However, if one wants to examine the inner workings of the hydraulic actuators or to study the actuator under different failure modes, the full dynamic model of the actuator should be used.

Since the control signal is $u_i(t) = k_{ha}k_v u_{ci}(t)$ under no fault condition, the relationship in the event of an actuator fault can be described as $u_i^f(t) = k_{ha}^f k_v u_{ci}(t)$. Thus, the ratio between $u_i(t)$ and $u_i^f(t)$ is essentially the effectiveness of one actuator. Define a variable l_i as

$$u_i^f(t) = l_i \cdot u_i(t), \qquad (2.3)$$

where $u_i^f(t) = l_i \cdot u_i(t)$ and $0 \le l_i \le 1$ is known as the actuator effectiveness. For a healthy actuator, $l_i = 1$. If the actuator suffers a complete failure (outage), then $l_i = 0$. A partial loss in the actuator effectiveness can be represented by $0 < l_i < 1$.

For simplicity, only a single control surface is considered in the above analysis. In practice, there are several control surfaces in an aircraft. These independently controlled surfaces form essential redundancies needed for fault tolerance, and together they generate the required moments to control the motion of the aircraft.

2.2.3   Modeling of Faults in Multiple Actuators

By extending the analysis of a single actuator, fault models for multiple independent actuators can be derived. In case of m actuators, the effectiveness factor l_i can be denoted as a diagonal element in the diagonal matrix $L = diag\{l_1, \ldots, l_m\}$ for i^{th} actuator. When potential failures in all m actuators are considered, the matrix L can be used to describe the effectiveness in any of them. In fact, such modeling techniques have been used in the literature [9, 20, 21]. With the diagonal matrix $L = diag\{l_1, \ldots, l_m\}$, the effectiveness of m independent actuators can be described as:

$$u^f(t) = Lu(t), \qquad (2.4)$$

where the diagonal element $0 \le l_i \le 1$, $(i = 1, \ldots, m)$ represents the effectiveness for i^{th} control channel, $u = [u_1, u_2, \ldots, u_m]^T$ is the desired control input vector, and $u^f = \left[u_1^f, u_2^f, \ldots, u_m^f\right]^T$ denotes the actual control actions exerted on the aircraft.

Consider the model of an aircraft that has been linearized at a desired operating condition described by:

$$\begin{cases} \dot{x}(t) = Ax(t) + Bu(t) + G\omega(t) \\ y(t) = Cx(t) \end{cases}, \tag{2.5}$$

where $x(t) \in \Re^n$ is the state vector, $\omega(t) \in \Re^r$ models a bounded external disturbance, $u(t) \in \Re^m$ and $y(t) \in \Re^p$ denote the control surface deflection and the measurement output vector, respectively. A, B, G, and C are known matrices with appropriate dimensions.

The system with actuator faults modeled in terms of the control effectiveness matrix can then be written as:

$$\begin{cases} \dot{x}(t) = Ax(t) + B_f u(t) + G\omega(t) \\ y(t) = Cx(t) \end{cases}, \tag{2.6}$$

where the post-fault control input matrix can be represented by $B_f = BL$.

To ensure acceptable control performance in the presence of actuator faults, an FTCS has to counteract the effects of the faults by utilizing the existing physical redundancies in the system. The following definition of actuator redundancy is restated.

Definition 2.1. *For a dynamic system described as Eq. (2.5), it is said to have $(m - p)$ degrees of actuator redundancy if the pair (A, b_i) is completely controllable $\forall i$ $(1 \leq i \leq m)$, where b_i is the ith column of control input matrix B, and the number of independent control inputs is more than the number of system outputs being controlled.*

2.3   Objectives and Formulation of Hybrid FTCS

The objective of the hybrid FTCS in this chapter can be stated as follows:

1. When a fault is detected, but not yet completely diagnosed, the performance deterioration of the closed-loop system should be slowed down through a passive control system; and

2. After the fault diagnosis is completed, the optimal performance can then be achieved through an appropriate reconfigurable controller.

In order to eliminate the steady-state tracking error for a step change maneuver and to deal with step disturbance, such as vertical wind gust, a controller with proportional and integral architecture [21] is used. The corresponding augmented system can be represented:

$$
\begin{cases}
\begin{bmatrix} e(t) \\ \dot{x}(t) \end{bmatrix} = \begin{bmatrix} 0 & -C \\ 0 & A \end{bmatrix} \begin{bmatrix} \int e(t)dt \\ x(t) \end{bmatrix} + \begin{bmatrix} 0 \\ B \end{bmatrix} u(t) + \begin{bmatrix} I & 0 \\ 0 & G \end{bmatrix} \begin{bmatrix} r(t) \\ \omega(t) \end{bmatrix} \\
\begin{bmatrix} \int e(t)dt \\ y(t) \end{bmatrix} = \begin{bmatrix} I & 0 \\ 0 & C \end{bmatrix} \begin{bmatrix} \int e(t)dt \\ x(t) \end{bmatrix}
\end{cases}
$$

$$(2.7)$$

where $e(t) = r(t) - y(t)$. Define the augmented state vector $x_a(t) = \left[\left(\int e(t) dt \right)^T, x^T(t) \right]^T$, the augmented external disturbance vector $\omega_a(t) = \left[r^T(t), \omega^T(t) \right]^T$, and the augmented measured output vector $y_a(t) = \left[\left(\int e(t) dt \right)^T, y^T(t) \right]^T$. Since the vertical gust considered is of a step form, the integral action of the controller can effectively compensate its effects.

Consequently, Eq. (2.7) can be rewritten as:

$$
\begin{cases}
\dot{x}_a(t) = A_a x_a(t) + B_a u(t) + G_a \omega_a(t) \\
y_a(t) = C_a x_a(t)
\end{cases}
$$

$$(2.8)$$

where $A_a = \begin{bmatrix} 0 & -C \\ 0 & A \end{bmatrix} \in \Re^{(l+n)\times(l+n)}$, $B_a = \begin{bmatrix} 0 \\ B \end{bmatrix} \in \Re^{(l+n)\times m}$, $G_a = \begin{bmatrix} I & 0 \\ 0 & G \end{bmatrix} \in \Re^{(l+n)\times(l+r)}$, and $C_a = \begin{bmatrix} I & 0 \\ 0 & C \end{bmatrix} \in \Re^{(l+p)\times(l+n)}$. Using the control effectiveness representation in Eq. (2.4), the augmented system with potential actuator faults can be described as:

$$
\begin{cases}
\dot{x}_a(t) = A_a x_a(t) + B_a L u(t) + G_a \omega_a(t) \\
y_a(t) = C_a x_a(t)
\end{cases}
$$

$$(2.9)$$

The general structure of the proposed hybrid FTCS is shown in Fig. 2.2. The system comprises an FDD scheme, decision-making mechanism, a normal controller, a passive FTCS, a reconfigurable mechanism, and a reconfigurable controller. In this chapter, only the controller design aspect is addressed. If a fault occurs in the actuator, the switching decision from the passive FTCS to the reconfigurable controller will be dependent on the convergence condition of the FDD scheme. An illustrative event diagram is given in Fig. 2.3, T_f, T_{det}, T_{dia}, and T_{rec} are the time of fault occurrence, fault detection, diagnosis, and controller reconfiguration, respectively. The FDD schemes require a certain amount of data to detect the fault and estimate the fault parameter in terms of loss of effectiveness. In the developed FTCS, once a failure is detected, the passive FTCS will be switched in first to slow down the performance deterioration. One can, therefore, "buy some time" by introducing the passive FTCS to let the FDD schemes obtain more accurate fault information for controller reconfiguration purpose at the second stage. A significant advantage of the

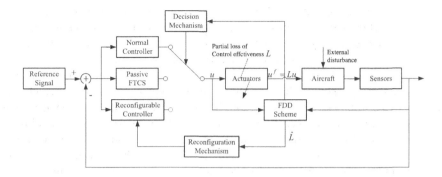

FIGURE 2.2: An illustrative diagram of the proposed FTCS.

proposed FTCS is to take this "intermediate" step to ease off the urgency before a reconfigurable controller becomes ready to be deployed. As more accurate diagnosis information becomes available, the reconfigurable controller can then be synthesized and subsequently switched in to improve the system performance under the circumstance.

Model-based FDD methods can be mainly classified into two types, parameter estimation and state estimation. As mentioned in Ref. [29], the convergence rate of a parameter estimation based FDD scheme is proportional to $1/k$, where k represents the number of samples taken for FDD purpose. On the other hand, in a state estimation based FDD method, the rate of convergence of the state estimation error is exponential with respect to the number of samples. The detailed information on the convergence rate of the parameter estimation method can be found in Ref. [29]. For the interest of space, it is omitted herein. In this chapter, the number of samples k is closely related to the estimated convergence of the fault estimation scheme. For instance in a parameter estimation based FDD unit, when the convergence accuracy is set to be 99.9%, the corresponding number of samples should be $1/(1-0.999)=1000$.

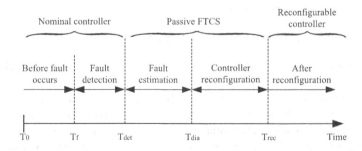

FIGURE 2.3: The event timing diagram of the proposed hybrid FTCS.

Assuming the sample interval is 0.02 sec, the fault estimation time can be calculated as $1000 \times 0.02 = 20$ sec. The beauty of the proposed scheme is that one can operate the system with the passive FTCS for as long as necessary until desired convergence of the fault estimation is achieved.

No FTCS can deal with all possible faults. The first condition is that the system has to possess sufficient degree of physical redundancies, which can be used to compensate the effects of the faults. The second condition is that a passive FTCS can be synthesized for a set of considered actuator failures. The switching function and the intervals between switching actions can also impact on the stability of the system [9]. The switching actions may also induce undesirable transients. In the proposed scheme, the stability can be maintained and switching transients can be minimized simply by limiting the number of switching activities by making use of a passive FTCS in transition before a more permanent reconfigurable controller is ready.

2.4   Design of the Hybrid FTCS

The methodologies of synthesizing the proposed FTCS are given herein. First, the structure of the closed-loop system with the tracking controller is presented. Sufficient conditions are provided in Lemma 2.1 and Theorems 2.1 and 2.2 for synthesizing the fault-tolerant controllers.

Consider the nominal system Eq. (2.8) with the following state feedback controller:

$$u(t) = K x_a(t) = K_e \int_0^t e(\tau) \, d\tau + K_x x(t), \qquad (2.10)$$

where $K = \begin{bmatrix} K_e, & K_x \end{bmatrix} \in \Re^{m \times (r+n)}$. The closed-loop system can be written as

$$\dot{x}_a(t) = (A_a + B_a K) x_a(t) + G_a \omega_a(t). \qquad (2.11)$$

The closed-loop system with potential loss of effectiveness in control channels can then be represented as

$$\dot{x}_a(t) = (A_a + B_a L K) x_a(t) + G_a \omega_a(t), \qquad (2.12)$$

where the matrix $L = diag\{l_1, \ldots, l_m\}$ describes the effectiveness of the actuators.

The controller design procedure can be divided into two steps. In the first step, a passive FTCS is determined as a stabilizing controller minimizing the upper bound of an associated quadratic cost function. In the second step, a reconfigurable controller is synthesized to further reduce the bound for the same cost function.

Define the following linear-quadratic (LQ) cost function:

$$J = \int_0^t \left(x_a^T (t) \, Q x_a (t) + u^T (t) \, R u (t) \right) dt, \tag{2.13}$$

where Q and R are symmetric positive semi-definite and positive definite weighting matrices, respectively. This performance index is chosen based on the fact that they are commonly used in aircraft controller design processes [9, 21, 73].

2.4.1   Passive FTCS Design Procedure

Assume that the linearized aircraft equation is given as follows:

$$\begin{cases} \dot{x}(t) = Ax(t) + Bu(t) + G\omega(t) \\ y(t) = Cx(t) \end{cases}, \tag{2.14}$$

where $x = \begin{bmatrix} \alpha & \beta & p & q & r \end{bmatrix}^T \in \Re^5$ consists of angle of attack (AOA), sideslip angle, roll rate, pitch rate, and yaw rate, $u = \begin{bmatrix} \delta_{rc} & \delta_{lc} & \delta_{roe} & \delta_{rie} & \delta_{lie} & \delta_{loe} & \delta_r \end{bmatrix}^T \in \Re^7$ denotes the deflections of right canard (RC), left canard (LC), right outer elevon (ROE), right inner elevon (RIE), left inner elevon (LIE), left outer elevon (LOE), and rudder (RUD), $\omega \in \Re^1$, represents the external disturbance, and $y = \alpha \in \Re^1$ is the output being controlled. From the redundancy definition [20], as far as the AOA is concerned, this airplane has 6 degrees of actuator redundancy.

From a practical perspective, different control inputs can have different effects on the system outputs. The deflections of control surfaces in an aircraft are all range limited. This means that a control surface that has a weak influence on the controlled output may easily become saturated, if it is used for what was not intended for. Specifically, the control of the yaw angle is primarily done through the RUD. However, the RUD plays a negligible role in an AOA maneuver. In this chapter, since only an AOA maneuver is considered, the RUD is not eligible to be considered as a redundant actuator. However, the canards and the elevons constitute a set of redundant actuators as far as the AOA is concerned.

According to Lemma 2.1 in Ref. [21], a sufficient condition for synthesizing a passive FTCS with the constraints of LQ performance and the closed-loop H_∞ norm against actuator failures can be restated as:

$$(A_a + B_a LK)^T P + P (A_a + B_a LK) + (1/\gamma^2) \, P G_a G_a^T P + K^T L^T R L K + Q < 0, \tag{2.15}$$

where γ corresponds to the H_∞ norm of the transfer function $T_{zw}(s)$ from the bounded exogenous disturbance $\omega_a(t)$ to the performance output $z(t)$ as defined in Eq. (2.16), which is commonly used to measure the disturbance attenuation in flight control system design [21]. The performance output $z(t)$

is defined as:

$$z(t) = \left[Q^{1/2}, 0\right]^T x_a(t) + \left[0, R^{1/2}\right]^T Lu(t). \tag{2.16}$$

It is worthy to note that $x_a^T \left(K^T L^T RLK + Q\right) x_a = z^T z$. Inequality (2.15) can be rewritten in the form:

$$\Omega_0 + \sum_{i=1}^{m} l_i \Omega_i + \sum_{i=1}^{m} l_i^2 \Theta_i < 0, \tag{2.17}$$

where $\Omega_0 = A_a^T P + P A_a + \left(1/\gamma^2\right) G_a^T G_a + Q$, $\Omega_i = K^T B_{ai}^T P + P B_{ai} K$, $B_{ai} = \left[\begin{array}{ccccc} 0 \ldots 0 & b_{ai} & 0 \ldots 0 \end{array} \right]$, $B_a = [b_{a1}, b_{a2}, \ldots, b_{am}]$, and the matrices $\Theta_i = K^T R_i K$, $R_i = diag\{0, \ldots, 0, r_i, 0, \ldots, 0\}$, r_i is the ith diagonal element of the positive definite matrix $R = diag\{r_1, r_2, \ldots, r_m\}$, m denotes the number of independent control surfaces.

Since the weighting matrix R is positive definite and $0 \leq l_i \leq 1$, the following condition holds:

$$\sum_{i=1}^{m} \Theta_i > \sum_{i=1}^{m} l_i^2 \Theta_i. \tag{2.18}$$

Therefore, the sufficient condition of satisfying (2.17) is that

$$\Omega_0 + \sum_{i=1}^{m} l_i \Omega_i + \sum_{i=1}^{m} \Theta_i < 0 \tag{2.19}$$

holds.

It can be seen that the matrix inequality (2.19) depends affinely on l_i, where l_i is the ith element of the diagonal matrix L. Therefore, the condition whether or not (2.17) holds for any value of l_i, $(0 \leq l_i \leq 1)$ can be examined by using only the bounded points $l_i = 0$ and $l_i = 1$, where $l_i = 0$ presents the worst case in the ith actuator.

The significance of the above result in practice is that it is possible to design a passive FTCS to slow down the rapid deterioration in system performance soon after the fault occurrence. The passive FTCS can accommodate the type of actuator failures with sufficient degrees of actuator redundancies. This concept is proposed and proved in Refs. [9, 20]. Since the fault can progress rapidly and the critical system states can even go beyond the system safety boundaries, the passive FTCS should be introduced as soon as possible to maintain the stability of the overall system as well as slow down the performance deterioration. The passive FTCS may not work any conceivable faults, however, for the faults as a loss of effectiveness defined in the current work, it works well. In the passive FTCS design process, it is assumed that if the designed passive FTCS can guarantee the stability of the closed-loop system in the worst case, any fault resulting in control effectiveness loss can also be accommodated. The possible failure modes exclude the case that all the actuators are completely failed, because at least one canard or elevon should be functional to ensure the AOA can be controlled. This complies with the redundancy requirements in fault-tolerant control design requirements [20]. The extreme case is that only one actuator is operational. The objective of the passive FTCS is to maintain the system stability under all the considered failures. In the current work, a passive FTCS is synthesized under the nominal and the extreme design basis faults, the control effectiveness matrices are selected as $L_0 = diag\{1, 1, 1, 1, 1, 1, 1\}$, $L_1 = diag\{1, 0, 0, 0, 0, 0, 0\}$, $L_2 = diag\{0, 1, 0, 0, 0, 0, 0\}$, $L_3 = diag\{0, 0, 1, 0, 0, 0, 0\}$, $L_4 = diag\{0, 0, 0, 1, 0, 0, 0\}$, $L_5 = diag\{0, 0, 0, 0, 1, 0, 0\}$, and $L_6 = diag\{0, 0, 0, 0, 0, 1, 0\}$.

In the design process, the Schur complement [74] is used to represent the sufficient condition for designing FTC in the form of an LMI [9, 21]. Commonly used performance indices, e.g., LQ index, the closed-loop H_∞ norm, and the following defined indices Eqs. (2.21) and (2.22) can be used to measure the conservatism of the designed controller.

In order to measure the conservatism of the FTCSs, the following performance measures are used in this chapter:

$$e_{perf}(t) = \|r(t) - y_{ftc}(t)\|_2, \qquad (2.20)$$

where $r(t)$ is the desired reference signal, while $y_{ftc}(t)$ denotes the closed-loop system output with either a passive or an active FTCS in the control loop. In addition, the mean and the maximum value of $e_{perf}(t)$, $t \in [t_1, t_2]$, as defined in Eqs. (2.21) and (2.22), are also used as the overall performance measures.

$$\bar{e}_{perf} = \frac{1}{t_2 - t_1} \int_{t_1}^{t_2} e_{perf}(t)\, dt, \qquad (2.21)$$

$$e_{perf\text{-max}} = \max_{t_1 \le t \le t_2} \{e_{perf}(t)\}, \qquad (2.22)$$

where $[t_1, t_2]$ covers the transition interval from the normal system operation to failure occurrence, and subsequently also include the period during the controller reconfiguration process.

Lemma 2.1. *(Reciprocal Projection Lemma) [75] Let P be any given positive definite matrix. The following statements are equivalent:*

(i)

$$\psi + \chi + \chi^T < 0, \qquad (2.23)$$

(ii) the LMI problem

$$\begin{bmatrix} \psi + P - (W + W^T) & \chi^T + W^T \\ \chi + W & -P \end{bmatrix} < 0, \qquad (2.24)$$

is feasible with respect to W, where ψ and χ are a symmetric matrix and a general matrix with appropriate dimensions.

Remark 2.1. *The conditions (i) and (ii) are equivalent; however the slack variable can provide additional flexibility. The passive FTCS design procedure employs the idea of the Reciprocal Projection Lemma to utilize this extra degree of freedom to choose the controller parameters. The idea of Lemma 2.1 and parameter-dependent Lyapunov functions are employed to derive the following theorem, which is used to synthesize the passive FTCS.*

Theorem 2.1. *Consider the closed-loop system given in Eq. (2.12). For a given scalar $\gamma > 0$ for all nonzero $\omega_a(t) \in L_2[0, \infty)$, Q and R are the weighting matrices used in LQ index, if there exist symmetric positive definite matrices $X_i = X_i^T > 0$ and any appropriately dimensioned matrices V, S_i, and N such that the following LMIs hold:*

$$\Pi_i = \begin{bmatrix} -(V+V^T) & \begin{matrix} V^T A_a^T + \\ N^T L_i^T B_a^T + X_i \end{matrix} & V^T & S_i^T G_a & V^T & N^T \\ * & -X_i & 0 & 0 & 0 & 0 \\ * & * & -X_i & 0 & 0 & 0 \\ * & * & * & -\gamma^2 I & 0 & 0 \\ * & * & * & * & -Q^{-1} & 0 \\ * & * & * & * & * & -R^{-1} \end{bmatrix} < 0,$$

(2.25)

*where the symbol * stands for a symmetric entry. The closed-loop system Eq. (2.12) obtains the upper bounds of performance indices of each considered scenarios,*

$$J_i < x_a^T(0) X_i^{-1} x_a(0) + \gamma^2 \int_0^t \omega_a^T \omega_a dt, \quad (i = 0, 1, \ldots, 6). \qquad (2.26)$$

The gain matrix for a passive FTC can be calculated as

$$K_{rel} = N_{opt} V_{opt}^{-1}, \qquad (2.27)$$

where N_{opt} and V_{opt} are the optimal solutions to inequality (2.25).

Proof. According to the sufficient condition in Ref. [21],

$$(A_a + B_a L_i K)^T P_i + P_i(A_a + B_a L_i K) + (1/\gamma^2) P_i G_a G_a^T P_i + K^T L_i^T R L_i K + Q < 0, \qquad (2.28)$$

where $i = 0, 1, \ldots, 6$. L_0 stands for the normal operation and $L_i, (i \neq 0)$ represents the ith design basis fault. Since the weighting matrix R is positive definite and $0 \leq l_i \leq 1$, the sufficient condition for satisfying (2.25) is

$$(A_a + B_a L_i K)^T P_i + P_i(A_a + B_a L_i K) + (1/\gamma^2) P_i G_a G_a^T P_i + K^T R K + Q < 0, \qquad (2.29)$$

where $i = 0, 1, \ldots, 6$. According to Lemma 2.1 above, by choosing $\psi = Q + (1/\gamma^2) P_i G_a G_a^T P_i + K^T R K$ and $\chi = P_i(A_a + B_a L_i K)$, the inequality (2.29) can also be rewritten by using the Schur complement and Lemma 2.1 as:

$$\begin{bmatrix} \begin{matrix} Q + (1/\gamma^2) P_i G_a G_a^T P_i + \\ K^T R K + P_i - (W + W^T) \\ P_i(A_a + B_a L_i K) + W \end{matrix} & (A_a + B_a L_i K)^T P_i + W^T \\ & -P_i \end{bmatrix} < 0. \quad (2.30)$$

The following inequality can be obtained by performing a congruence transformation

$$\begin{bmatrix} V & 0 \\ 0 & X_i \end{bmatrix} \text{ with } V = W^{-1}, X_i = P_i^{-1}, \text{ and } X_i = X_i^T.$$

in (2.30). The derivation process is as:

$$\begin{bmatrix} V^T & 0 \\ 0 & X_i \end{bmatrix} \begin{bmatrix} Q + (1/\gamma^2) P_i G_a G_a^T P_i + \\ K^T RK + P_i - (W + W^T) & (A_a + B_a L_i K)^T P_i + W^T \\ P_i (A_a + B_a L_i K) + W & -P_i \end{bmatrix} \begin{bmatrix} V & 0 \\ 0 & X_i \end{bmatrix} \tag{2.31}$$

$$= \begin{bmatrix} V^T QV + (1/\gamma^2) V^T P_i G_a G_a^T P_i V + \\ V^T K^T RKV + V^T P_i V - (V + V^T) & V^T (A_a + B_a L_i K)^T + X_i \\ (A_a + B_a L_i K) V + X_i & -X_i \end{bmatrix}. \tag{2.32}$$

Let $S_i = P_i V$, $N = KV$, Eq. (2.32) can be represented as (2.33),

$$\begin{bmatrix} -(V + V^T) & V^T A_a^T + N^T L_i^T B_a^T + X_i & V^T & S_i^T G_a & V^T & N^T \\ * & -X_i & 0 & 0 & 0 & 0 \\ * & * & -X_i & 0 & 0 & 0 \\ * & * & * & -\gamma^2 I & 0 & 0 \\ * & * & * & * & -Q^{-1} & 0 \\ * & * & * & * & * & -R^{-1} \end{bmatrix} < 0. \tag{2.33}$$

Hence, the controller $u(t) = K x_a(t)$ stabilizes the augmented system Eq. (2.12). Furthermore, substituting $u(t) = K x_a(t)$ into the performance index Eq. (2.13) yields:

$$J_i = \int_0^t x_a^T \left(Q + K^T RK \right) x_a dt$$

$$< -\int_0^t x_a^T \left[(A_a + B_a L_i K)^T P_i + P_i (A_a + B_a L_i K) + (1/\gamma^2) P_i G_a G_a^T P_i \right] x_a dt$$

$$= -\int_0^t \left[(\dot{x}_a - G_a \omega_a)^T P_i x_a + x_a^T P_i (\dot{x}_a - G_a \omega_a) + (1/\gamma^2) x_a^T P_i G_a G_a^T P_i x_a \right] dt$$

$$\leq -\int_0^t d \left(x_a^T P_i x_a \right) + \gamma^2 \int_0^t \omega_a^T \omega_a dt$$

$$\leq x_a^T (0) P_i x_a (0) + \gamma^2 \int_0^t \omega_a^T \omega_a dt = x_a^T (0) X_i^{-1} x_a (0) + \gamma^2 \int_0^t \omega_a^T \omega_a dt.$$

The subscript i ($i = 0, 1, \ldots, 6$) represents the scenarios considered in the passive FTCS design, and J_0 stands for the LQ performance of the normal system. □

Remark 2.2. *Theorem 2.1 provides a sufficient condition for the existence of the passive FTCS in the presence of actuator failures. In the proposed approach, each design basis fault corresponds to a Lyapunov matrix and the slack variable V eliminates the product terms between the Lyapunov matrix X_i and the system matrices. Compared to the sufficient condition given in Ref. [9], Theorem 2.1 in terms of the parameter-dependent Lyapunov functions and the slack variable V has extended the freedom of the design method.*

Consequently, the passive FTCS design procedure can be formulated to minimize the following objective:

$$\min Trace\,(Y_0)\,, \qquad (2.34)$$

subject to (2.25), and

$$\begin{bmatrix} Y_i & I \\ I & X_i \end{bmatrix} > 0, \quad (i = 0, 1, \dots, 6)\,. \qquad (2.35)$$

From Refs. [9, 21], the upper cost bounds $E\left(x_0^T P_i x_0\right) = Trace\,(P_i) = Trace\left(X_i^{-1}\right)$, where the notation E denotes the expectation operator with respect to the initial state x_0, which is assumed to be a zero-mean random variable with a covariance matrix $E\left(x_0^T x_0\right)$. The LMI representations (2.35) in terms of Y_i and X_i are used to transform the objective function $Trace\,(P_i) = Trace\left(X_i^{-1}\right)$ to a linear function $Trace(Y_i)$. The performance in the nominal case can be optimized via (2.34). $Trace(Y_0)$ is selected as the objective function in the optimization because the nominal system operation still covers the majority fraction of the operational life of the system in practice. Other indices $J_i\,(i = 1, \dots, 6)$ can be limited by constraining $Trace(P_i)$. Through the above optimization process, the gain of the passive FTC can be obtained by $K_{rel} = N_{opt}(V_{opt})^{-1}$, where N_{opt}, V_{opt} denote the optimal solution of the passive FTCS design procedure.

2.4.2   Reconfigurable Controller Design Procedure

Denote $L = \hat{L}$, where $\hat{L}$ is assumed to be the correctly estimated control effectiveness matrix after convergence of the FDD scheme. The corresponding closed-loop system can be described as,

$$\dot{x}_a\,(t) = \left(A_a + B_a \hat{L} K\right) x_a\,(t) + G_a \omega_a\,(t)\,. \qquad (2.36)$$

Therefore, the reconfigurable controller can be synthesized as follows.

Theorem 2.2. *Consider the closed-loop system in Eq. (2.36). For a given scalar $\gamma > 0$ for all nonzero $\omega_a\,(t) \in L_2\,[0\,,\infty)$, Q and R are the weighting matrices used in the LQ index, if there exist a symmetric positive definite* *matrix $X_{rec} = X_{rec}^T > 0$ and any appropriately dimensioned matrices V_{rec}, S_{rec}, and N_{rec} such that the LMI holds:*

$$
\Pi_{rec} =
\begin{bmatrix}
-(V_{rec}+V_{rec}^T) & \begin{matrix} V_{rec}^T A_a^T + \\ N_{rec}^T \hat{L} B_a^T + X_{rec} \\ -X_{rec} \end{matrix} & V_{rec}^T & S_{rec}^T G_a & V_{rec}^T & N_{rec}^T \hat{L}^T \\
* & -X_{rec} & 0 & 0 & 0 & 0 \\
* & * & -X_{rec} & 0 & 0 & 0 \\
* & * & * & -\gamma^2 I & 0 & 0 \\
* & * & * & * & -Q^{-1} & 0 \\
* & * & * & * & * & -R^{-1}
\end{bmatrix}
< 0,
$$

$$(2.37)$$

then the closed-loop system Eq. (2.36) obtains the upper bounds of the performance index,

$$
J < x_a^T(0) X^{-1} x_a(0) + \gamma^2 \int_0^t \omega_a^T \omega_a \, dt. \tag{2.38}
$$

The reconfigurable controller can be solved by

$$
K_{rec} = N_{rec\_opt} V_{rec\_opt}^{-1}. \tag{2.39}
$$

Proof. The proof procedure is similar to that in Theorem 2.1.

When the loss of effectiveness matrix $\hat{L}$ estimated by the FDD unit is available, the reconfigurable controller can be synthesized through the following convex optimization process. The objective is to minimize the following quantity:

$$
\min Trace\,(Y_{rec}), \tag{2.40}
$$

subject to (2.37), and

$$
\begin{bmatrix} Y_{rec} & I \\ I & X_{rec} \end{bmatrix} > 0. \tag{2.41}
$$
$\square$

Remark 2.3. *In this case, the reconfigurable controller can be determined to minimize the condition $Trace(Y_{rec})$ following a similar procedure to that described in the passive FTCS design. When the sufficient condition (2.37) of the existing controller with the constraint of (2.40) satisfies, the reconfigurable controller can be synthesized by the above optimization process. The optimization objective is to minimize the quadratic index when the exact fault information is available.*

2.4.3 Switching Function among Different Controllers

When switching from one controller to another on-line, if there are any mismatches in values at the respective controller outputs, the switching will induce transients. Therefore, to minimize these transients, it is important to ensure that the difference between the two corresponding controller outputs is as small as possible at the instant of switching. In the current work, a smooth switching function is used.

Assume that K_{nor} is to be replaced by K_{rel} at t_0, hence represented as $K_{nor}(t_0)$ and $K_{rel}(t_0)$, the following switching function is used:

$$K_{rel}(t) = K_{rel}(t_0) + (K_{nor}(t_0) - K_{rel}(t_0)) \, e^{-\tau(t-t_0)}, \qquad (2.42)$$

where τ depends on the dynamics of the closed-loop system. This switching process guarantees a smooth transition from K_{nor} to K_{rel}. The same switching function has also been used for switching from K_{rel} to K_{rec}.

2.5   Numerical Case Studies

2.5.1   Description of the Aircraft

The ADMIRE (Aero-Data Model in Research Environment) benchmark aircraft [76] with inherent actuator redundancies is selected for the current case studies. The states, control inputs, outputs, and external disturbance with a magnitude of $2\,\text{m/s}$ that lasts 100 sec in the simulation are shown in Eq. (2.14) and its following description.

The trimming values of the aircraft are: $Ma = 0.45$, $H = 3000\,\text{m}$, $V_t = 147.86\,\text{m/s}$, $\alpha = 3.737\,43\,°$, $\beta = 0$, $\delta_{rc} = \delta_{lc} = -0.0518\,°$, $\delta_{roe} = \delta_{rie} = \delta_{lie} = \delta_{roe} = -0.036\,18\,°$, $\delta_r = 0$.

The nominal matrices A, B, G, and C under the given trimming condition are

$$A = \begin{bmatrix} -1.0649 & 0.0034 & -0.0000 & 0.9728 & 0.0000 \\ 0.0000 & -0.2492 & 0.0656 & -0.0000 & -0.9879 \\ 0.0000 & -22.5462 & -2.0457 & -0.0000 & 0.5432 \\ 8.1633 & -0.0057 & -0.0000 & -1.0478 & 0.0000 \\ 0.0000 & 1.7970 & -0.1096 & 0.0000 & -0.4357 \end{bmatrix},$$

$$B = \begin{bmatrix} -0.0062 & -0.0062 & -0.0709 & -0.1172 & -0.1172 & -0.0709 & 0.0003 \\ -0.0072 & 0.0072 & 0.0039 & 0.0188 & -0.0188 & -0.0039 & 0.0627 \\ 1.2456 & -1.2456 & -10.6058 & -9.2345 & 9.2345 & 10.6058 & 5.3223 \\ 2.7172 & 2.7172 & -2.4724 & -4.0101 & -4.0101 & -2.4724 & 0.0108 \\ -0.7497 & 0.7497 & -0.4923 & -1.1415 & 1.1415 & 0.4923 & -3.7367 \end{bmatrix},$$

$$G = \begin{bmatrix} -0.0072 & 0.0000 & 0.0000 & 0.0551 & -0.0000 \end{bmatrix}^T, C = [1,0,0,0,0].$$

The design parameters for passive FTCS and the reconfigurable controller are: $\gamma_{pass} = 2$, $\gamma_{rec} = 2$, $Q = diag\{10, 1, 1, 1, 1, 1\}$, $R = diag\{1, 1, 1, 1, 1, 1\}$. In the following studies, the reference for AOA is a pulse of 2 deg in magnitude, and 80 sec duration starting at 20 sec. The time of fault occurrence is assumed to be 40 sec. The reconfigurable controller is commissioned at 45 sec.

2.5.2   Performance Evaluation under the Passive FTCS

Two fault scenarios have been considered. In the first case, two faults are assumed to have occurred and the control effectiveness matrix is set to be $L_{case1} = diag\{1, 1, 0.1, 0.1, 1, 1, 1\}$. In case 2, considerate is assumed that here are four faults and at the matrix becomes $L_{case2} = diag\{1, 1, 0.1, 0.1, 0.05, 0.05, 1\}$. The performance of the passive FTCS, which is synthesized with the consideration of the pre-assumed faults in Chapter 2.4, is compared with that of the active FTCS that is designed based on fault case 1. The responses of AOA with both the passive and active FTCSs in fault cases 1 and 2 are illustrated in Fig. 2.4. As shown in Fig. 2.4, the passive FTCS can deal with both cases 1 and 2. The performance of the active FTCS in case 1 is satisfactory. However, it cannot maintain the system stability in case 2. The performance indices defined in Eqs. (2.21) and (2.22) are tabulated in Table 2.2. Fig. 2.4 and Table 2.2 show that the passive FTCS is more robust against various fault scenarios, since the passive FTCS can maintain the stability for both cases 1 and 2, on the other hand, the active FTCS cannot accommodate the fault in case 2. Another important feature of the passive FTCS is to slow down the deterioration of the system performance so that the FDD can have extra time for correct fault estimation and control reconfiguration. This can be shown in Fig. 2.5 where the performance using a nominal controller is compared with that obtained with a passive FTCS in the event of actuator failures. The effective matrix of actuators is assumed to be $L_{buytime} = diag\{1, 1, 0.1, 0.1, 0.05, 0.05, 1\}$. In any practical system, the system output always has some permissible ranges, beyond which severe damage can happen. In the current study, the variable AOA is restricted within

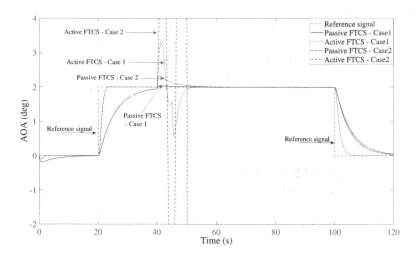

FIGURE 2.4: Performance comparison between passive and active FTCSs.

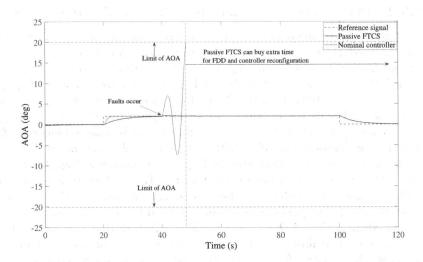

FIGURE 2.5: Comparison between nominal controller and passive FTCSs.

the range of $[-20°, 20°]$. By using a passive FTCS, it will take a lot longer for AOA to get out of this absolute boundary than in the case with a nominal controller. This situation can be observed in Fig. 2.5. Note that the time "bought" by the passive FTCS should be more than the time required for the convergence of the FDD plus the time for control reconfiguration. The amount of FDD time relies on the estimation method selected [29]. The normal controller has a faster response than that of the passive FTCS in fault free case. Since the passive FTCS is synthesized with consideration of both nominal and the worst case scenarios (only one actuator is left), the passive FTCS is less aggressive in its control effort. However, when faults occur, the passive FTCS can maintain the stability of the post-fault system. In an actuator failure situation, the most important concern is how to keep the system safe and stable, rather than achieving desirable performance as that in the normal case. This is exactly why a passive FTCS is used as the first controller in the transition process between the occurrence of a fault and the switching in of the active FTCS.

TABLE 2.2: Comparison of performance indices.

	Passive FTCS with Fault case 1	Active FTCS with Fault case 1	Passive FTCS with Fault case 2	Active FTCS with Fault case 2
$\bar{e}_{perf}$	0.0914	0.0822	0.0976	Oscillatory
e_{perf_max}	0.038	1.45	0.325	Oscillatory

2.5.3 Performance Evaluation under Reconfigurable Controller

Assuming that the actual control effectiveness matrix is set to be $L_{compare} = diag\{1, 1, 0.1, 0.1, 0.1, 0.1, 1\}$. The performance of the system under (1) A passive FTCS, (2) active FTCSs with incorrect FDD information, and (3) an active FTCS with correct FDD will be compared. In (1), a passive FTCS is designed with consideration of both normal and fault cases. For (2), two cases are considered for different convergence assumptions. In the case 1, it is assumed that the estimated effectiveness factors $\hat{i}_3 = \hat{i}_4 = \hat{i}_5 = \hat{i}_6 = 0.8$. In the case 2, the estimated effectiveness factors are supposed to be $\hat{i}_3 = \hat{i}_4 = 0.5$, $\hat{i}_5 = \hat{i}_6 = 0.1$. In scenarios (3) the reconfigurable controller is designed based on the converged fault estimation: $\hat{i}_3 = \hat{i}_4 = \hat{i}_5 = \hat{i}_6 = 0.1$. The corresponding responses of the closed-loop system in the above scenarios are shown as Fig. 2.6.

Between passive and active FTCSs, the performance of the passive FTCS is inferior to that of the active FTCS with the correct FDD information. This is because the passive FTCS has to deal with multiple fault cases in the design process. This can be seen clearly from Fig. 2.6. In fact, the accuracy of the FDD information has a profound impact on the performance of the active FTCS. As indicated in Fig. 2.6, with incorrect FDD information of case 1, the reconfigurable controller cannot always guarantee the stability of the post-fault system. With the improved convergence of FDD scheme, the active FTCS can not only stabilize the fault system but also achieve much better performance. The performance indices under different cases are shown

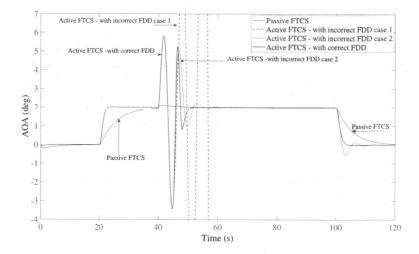

FIGURE 2.6: Comparison among passive FTCS, active FTCSs with correct and incorrect FDD.

in Table 2.3. The index e_{perf_max} in Table 2.3 with the active FTCS is larger than that with the passive FTCS, this is because the switch action in the active FTCS may cause unexpected transients. In summary, the indices achieved by passive FTCS are inferior to those obtained by active FTCS with correct FDD. Active FTCS with correct FDD achieves the best performance among the selected scenarios. It can be concluded that the system performance is not optimized in the passive FTCS. However, since various fault scenarios have to be taken into account at the design stage, the passive FTCS can accommodate a large kind of faults, which is "robust" to various fault situations. Conversely, the active FTCS, which focuses on the optimization of system performance based on the correct fault information, is less conservative than the passive FTCS. However, an incorrect FDD result will render the system completely unstable and could lead to a serious consequence.

The corresponding deflections of RC and LC that are still healthy in the fault scenario are indicated in Fig. 2.7. The deflections of the healthy actuators under the passive FTCS are larger than those under the active FTCS. However, from Fig. 2.6 and also numbers in Table 2.3, the performance of the AOA response with the passive FTCS is worse than that with the active FTCS. It can be readily concluded that, the passive FTCS uses more energy than the active FTCS, but the performance is inferior to that of the active control. This is another indication of the conservativeness of the passive FTCS.

2.5.4   Nonlinear Simulation of the Hybrid FTCS

The objective of the hybrid FTCS is to combine the merits of both passive and active FTCSs. The key contribution of this chapter is to stabilize the system first by a robust controller, while the FDD unit is carrying out the fault diagnosis tasks. Since the results from the linearized aircraft equation based on the small-perturbation assumption may not be general enough, the developed FTCS is validated using the fully nonlinear aircraft model. The fault scenario addressed herein is the loss of the control effectiveness in ROE, RIE, and LIE, the actual effectiveness factor is set to be 0.1 in the simulation studies. The passive FTCS is introduced 2.5 sec after the faults occur. In practice, it should be noted that the FDD delay depends on the nature of system failures and the selected FDD method [29]. In the current simulation,

TABLE 2.3: Comparison of performance indices.

	Passive FTCS	Active FTCS Incorrect FDD case 1	Active FTCS Incorrect FDD case 2	Active FTCS Correct FDD
$\bar{e}_{perf}$	0.0905	Oscillatory	0.1215	0.0852
e_{perf_max}	0.1	Oscillatory	3.06	3.25

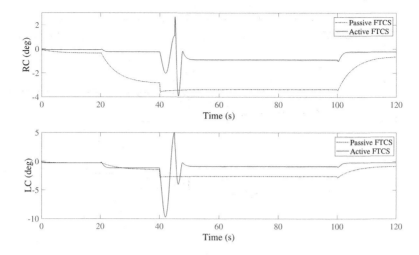

FIGURE 2.7: Comparisons of control surface deflections between active and passive FTCSs.

the assumed fault estimation time plus the controller reconfiguration time is assumed to be 17 sec, 20 sec, and 25 sec, respectively. To be more precise, in the simulation, the reconfigurable controller is switched in at 57 sec, 60 sec, and 65 sec, respectively. By setting various switching intervals in the nonlinear simulations, the sensitivity of the proposed FTCS scheme with respect to FDD time and reconfiguration time can be illustrated. The corresponding AOA responses and control surface deflections are shown in Fig. 2.8.

In the nonlinear simulations, the nominal, the passive FTC, and the active controllers with RIE, ROE, and LIE faults are:

$$K_{nor} = \begin{bmatrix} 0.8463 & -0.8696 & -0.0647 & -0.0572 & -0.3870 & 0.1867 \\ 0.8460 & -0.8696 & 0.0635 & 0.0573 & -0.3872 & -0.1872 \\ -1.0819 & 0.9965 & -0.4948 & 0.4629 & 0.3771 & 0.0869 \\ -1.7638 & 1.6223 & -0.5702 & 0.4013 & 0.6142 & 0.2534 \\ -1.7648 & 1.6230 & 0.5727 & -0.4014 & 0.6125 & -0.2527 \\ -1.0826 & 0.9971 & 0.4964 & -0.4630 & 0.3773 & -0.0865 \\ 0.0058 & -0.0045 & -0.5012 & -0.2429 & -0.0012 & 0.9452 \end{bmatrix},$$

$$K_{rel} = \begin{bmatrix} 1.5870 & -6.9194 & -1.2044 & -0.0467 & -2.2668 & 0.4225 \\ 1.5928 & -6.9401 & 1.2036 & 0.0466 & -2.2735 & -0.4216 \\ -1.8954 & 9.0172 & 0.3467 & 0.2061 & 2.9496 & -0.4701 \\ -3.0925 & 12.0253 & -0.4574 & 0.1882 & 3.8980 & 0.0051 \\ -3.0895 & 12.0127 & 0.4642 & -0.1877 & 3.8939 & -0.0084 \\ -1.8963 & 9.0206 & -0.3404 & -0.2060 & 2.9507 & 0.4672 \\ 0.0083 & -0.0265 & -6.9459 & 0.1264 & -0.0083 & 3.0709 \end{bmatrix},$$

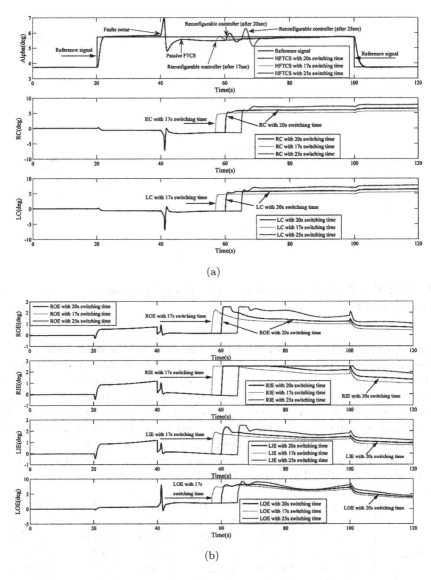

FIGURE 2.8: Nonlinear simulation with different switching intervals.

$$K_{rec} = \begin{bmatrix} 0.8463 & -0.8696 & -0.0647 & -0.0572 & -0.3870 & 0.1867 \\ 0.8460 & -0.8696 & 0.0635 & 0.0573 & -0.3872 & -0.1872 \\ -10.8188 & 9.9645 & -4.9482 & 4.6288 & 3.7707 & 0.8691 \\ -17.6376 & 16.2229 & -5.7017 & 4.0126 & 6.1245 & 2.5338 \\ -17.6482 & 16.2300 & 5.7271 & -4.0138 & 6.1252 & -2.5267 \\ -1.0826 & 0.9971 & 0.4964 & -0.4630 & 0.3773 & -0.0865 \\ 0.0058 & -0.0045 & -0.5012 & -0.2429 & -0.0012 & 0.9452 \end{bmatrix}.$$

As illustrated in Fig. 2.8, the passive FTCS is employed to stabilize the post-fault system while the FDD unit is perusing fault estimation. After the correct diagnostic information is available, the reconfigurable controller is synthesized and then commissioned to improve system performance. In the nonlinear simulations, the switching function Eq. (2.42) is used to prevent switching induced transients in the event of controller switching. It also can be seen that under different switching time intervals, the proposed FTCS performs satisfactorily in the nonlinear environment as well. The longer the switching time is, the larger the amount of control surface deflection becomes. From Fig. 2.8, the sensitivity of the proposed approach has been validated. The healthy control surfaces (RC, LC, and LOE) have to work more diligently than in the nominal case to track the reference commands.

It should be mentioned that an aircraft is a highly nonlinear system. No aircraft is controlled by a single linear controller. In a realistic flight control system, dozens of flight regions [76] are defined according to a number of selected nonlinear conditions, such as altitude, dynamic pressure, and airspeed, etc. There are several operating points within a flight region. Commonly, one linear controller is synthesized for each operating point. A series of linear controllers are used to maneuver an aircraft within and among regions based on the real-time information of altitude, dynamic pressure, and airspeed. In this chapter, the proposed FTCS is designed only for one operating condition within one region where the altitude and dynamic pressure do not change significantly. Hence, the nonlinear simulations are performed around this operating condition. If this design approach is used for the entire operating aircraft, the similar design process has to be carried out for each and every operating point.

2.6 Conclusions

A model of an aircraft with loss of control effectiveness is developed based on analysis of physical faults in the hydraulic driven control surfaces. A hybrid FTCS combining a passive and an active FTCS approaches is proposed to counteract partial actuator faults. The unique feature of the proposed system is that, in the presence of actuator faults, the passive FTCS can be used first to stabilize the system with minimal fault information. Once the more detailed fault diagnostic information becomes available, a reconfigurable controller can then be synthesized and used to improve the performance. The effectiveness of the proposed FTCS scheme and the design procedure are validated by using both linear and nonlinear case studies.

2.7 Notes

The main contribution of this chapter is first to examine the physical insights into actuator failures so that partial failures can be represented in terms of control channel effectiveness. After analyzing the degree of actuator redundancies in the system, a hybrid FTCS which combines passive and active FTCSs is proposed to counteract the partial failures in the actuators. The passive FTCS can guarantee the stability of the post-fault system and the active FTCS with the correct fault information can improve the performance. In the controller design process, the slack method that recasts common LMI characterizations as augmented LMI representations have been used to synthesize the controllers. The effectiveness of the proposed FTCS is demonstrated in a flight control system subject to loss of actuator effectiveness.

Chapter 3

Safety Control System Design against Control Surface Impairments

3.1　Introduction

A control surface is an essential component in any flight control system. It links control commands to physical actions exerted on the aircraft to achieve specific flight behavior. One of the unique situations encountered by military aircraft in combat is that the aircraft can be exposed to enemy fire, which may lead to control surface damage. During the Vietnam War, it is reported that more than 20% of aircraft losses are due to damage inflicted to hydraulic lines and control surfaces [77]. A commercial aircraft may also lose some of its control surfaces due to aging of structure, or hitting a flying object, such as a bird, etc. According to an NTSB (National Transportation Safety Board) study reported in Ref. [78], control surface impairments, mechanical, and/or hydraulic line failures resulted in a total of 98 fatal accidents in 2005. Damage to a control surface not only reduces the control effectiveness, but also alters the aerodynamics of the aircraft. The objective of this chapter is to present a new approach to accommodate control surface impairments by not only considering the loss of control effectiveness, but also the change of aerodynamic characteristics. A realistic and fully nonlinear aircraft model has been used in this study.

Among a large number of research works focused on the loss of control effectiveness, few approaches were developed in the presence of control surface impairments. In Ref. [79], after fault detection and surface impairment damage estimation, control allocation based on the pseudo-inverse method is used to counteract the effects caused by partial loss of a right horizontal stabilator. However, the change in system dynamics is not considered. Subject to this type of fault, reliable controllers are also used to cope with fault situations [21, 80, 81, 82]. An iterative linear matrix inequality (LMI) based approach that reduces the conservatism in reliable flight control system design is presented in Ref. [21]. A reliable controller design method against control surface impairments using quadratic parameter-dependent Lyapunov functions is provided in Ref. [80], An FTCS based on polytopic LPV systems is developed for multiple actuator failures in Ref. [83]. In the above references, although reliable controllers have been designed against control surface impairments [21, 80, 81, 82], however, a single reliable controller may not be able to cope with all potential control surface impairments since different faults in control surfaces can exhibit distinctive characteristics. Also, no active control strategy is used to counteract control surface impairments. In the current study, an LPV approach is adopted to describe the failure models of control surface impairments, and used to synthesize the fault-tolerant controllers. In comparison with the reliable controller design approach, one advantage of the proposed method is that the damaged area of the control surface is used as the scheduling parameter, which results in a less conservative control system under the specific fault situation.

In any FTCS design, redundancy analysis has to be performed first. The aim of redundancy analysis is to analyze the existence of solutions and explore the inherent interdependence among various control inputs. After that, one can judge whether the fault-tolerant controller (FTC) can functionally accommodate the fault situation or not. It is understandable that no controller can be designed to recover any uncontrollable or non-recoverable failures within the physical limitations on control surface movements and deflection rates. Moreover, redundancy analysis can also provide some guidance on the modification of control surface configuration in response to specific failures.

The rest of the chapter is organized as follows: A nonlinear aircraft model, control surface dynamics, and the corresponding linearized model are introduced in Chapter 3.2. The existence of a solution is examined through control surface redundancy analysis in Chapter 3.3, and the control problem of surface impairments is also formulated. In Chapter 3.4, a polytopic LPV model and an improved bounded real lemma (BRL) are proposed for FTC design. Simulation results are illustrated in Chapter 3.5 to validate the designed method, followed by some concluding remarks in Chapter 3.6. From the analysis and dynamic simulation, it is concluded that the proposed approach is effective in maintaining aircraft integrity under various control surface impairments.

Notation. *Throughout this chapter, the superscript T stands for matrix transposition, and the symbol $*$ within a matrix represents a symmetric entry. $\mathbf{R}^n$ denotes the n-dimensional Euclidean real space, $\mathbf{R}^{n \times m}$ is the set of all $n \times m$ real matrices. $\mathbf{C}^{\perp}$ denotes an orthogonal basis for the null space of $\mathbf{C}$.*

3.2   Aircraft Model with Redundant Control Surfaces

Even though the analysis and the design approaches are not limited to a specific type of aircraft in this work, however, it is advantageous to work with a specific aircraft system to explain the concepts and to validate the design procedure. We have selected the ADMIRE (Aero-Data Model in Research Environment) benchmark aircraft model [76], a generic model of a small single-seat fighter with a delta-canard configuration. In this aircraft model, there are twelve states and ten available control effectors. The ADMIRE aero-data are from Saab Aerospace, and the configuration of the model is based on the Gripen fourth-generation combat aircraft. The model is developed by the Swedish Defense Research Agency, as one of the GARTEUR (Group of Aeronautical Research and Technology in Europe) benchmark models in an Action Group AG-11 project for flight control clearance investigation [76]. The AD-MIRE benchmark aircraft has multiple physical redundancies, and this model has been used in several FTCS research studies.

The ADMIRE benchmark aircraft is shown in Fig. 3.1(a) whose airframe is referred from Ref. [76]. This figure also illustrates the control inputs, coordinate axes, and states. The available control effectors in ADMIRE are the left and right canards (LC, RC), left and right outer elevons (LOE, ROE), left and right inner elevons (LIE, RIE), rudder, as well as engine thrust, leading edge flaps, and landing gear. The states consist of angle of attack (AOA), sideslip angle, roll rate, pitch rate, yaw rate, and aircraft velocity along X, Y, and Z axes, respectively. For control design purposes, only the first seven control surfaces and the first five states are considered in the current investigation.

The aerodynamic characteristics of a control surface are expressed in terms of normal force, axial force, and moment around some fixed points or axes. Damage to a control surface, such as a torn off part of the controllable surface area, implies changes in the aerodynamic behavior of the aircraft. Such impairments can be effectively modeled with a parameter. The relationship between the control surface impairment and the parameter is illustrated in Fig. 3.1(b). The objective of the proposed active FTCS is to design a control system such that the integrity of the aircraft can be maintained under various degrees of control surface failure. To carry out the analysis and design process, a linearized fault model is derived from the nonlinear aircraft model.

3.2.1   Nonlinear Aircraft Model

The dynamics of ADMIRE benchmark aircraft are highly nonlinear. Under the rigid body assumption, the equations of motion can be derived as follows. The wind-axes force equations are

$$m\dot{V} = T\cos\alpha\cos\beta - D + G_{xa}, \tag{3.1}$$

$$mV\dot{\beta} = -T\cos\alpha\sin\beta + Y - mVr_w - G_{ya}, \tag{3.2}$$

$$mV\cos\beta\dot{\alpha} = q - T\sin\alpha - L + mVq_w + G_{za}. \tag{3.3}$$

The body-axes moment equations are

$$\dot{p} = (C_1 \cdot r + C_2 \cdot p) \cdot q + C_3 \cdot M_x + C_4 \cdot M_z, \tag{3.4}$$

$$\dot{q} = C_5 \cdot p \cdot r - C_6 \cdot (p^2 - r^2) + C_7 \cdot M_y, \tag{3.5}$$

$$\dot{r} = (C_8 \cdot p - C_2 \cdot r) \cdot q + C_4 \cdot M_x + C_9 \cdot M_z. \tag{3.6}$$

The definitions of the symbols and their physical meanings are explained in Appendix B.

3.2.2 Actuator Dynamics

Actuators represent the links between the control signals issued by the controller and the manipulated variables which exert physical actions onto the aircraft. Actuators in an aircraft are often in the form of hydraulically-driven control surfaces to provide the desired forces and moments to maneuver the aircraft. The canard actuators are modeled as a first order lag with a time constant of $0.05\,\text{s}$, rate limits of $50\,°/\text{s}$ and minimal and maximal deflection

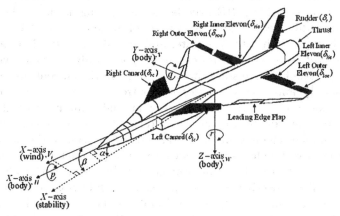

(a) Definition of states, reference frame, and control surface configuration.

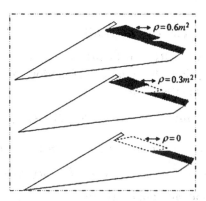

(b) Relation between faults and varying parameter.

FIGURE 3.1: Diagram of ADMIRE aircraft.

limits of $-55°$, and $25°$, respectively. The rudder actuator is modeled as a first order lag with a time constant of $0.05\,\mathrm{s}$, rate limits of $50°/\mathrm{s}$, and deflection limits of $30°$. The elevon actuators are also modeled as a first order lag with a time constant of $0.05\,\mathrm{s}$, rate limits of $50°/\mathrm{s}$, and deflection limits of $25°$. These are typical parameters adopted for the ADMIRE benchmark aircraft.

3.2.3   Linearized Aircraft Model with Consideration of Faults

Assumption 3.1. *The wing mean geometric chord, wing chord, and aerodynamic center are assumed to be normal when the control surface impairment occurs. The effect of every point on one single control surface is assumed to be the same.*

From Fig. 3.1(b), it can be seen that, when control surface impairment occurs, the effective area that is denoted by a parameter ρ will change. As a result, the coefficients of force and moment will also change since they are functions of the control surface area [73]. Due to this fact, control surface impairments will cause changes to the aerodynamic behavior of the aircraft. Therefore, the linearized system matrices will change after trimming. When control surface impairment occurs, not only the effectiveness of the control input is affected, but also the aerodynamics of the aircraft is changed.

From Eqs. (3.1) to (3.6), each fault model that is based on the small-perturbation assumption [73] as well as Assumption 3.1 can be obtained by linearizing the aircraft model after trimming under the corresponding fault conditions. Before using polytopic systems to describe the nominal and faulty aircraft models under the same initial altitude and total velocity, the relationship between the parameter ρ and the corresponding characteristic polynomial of the aircraft needs to be determined. Specifically, the characteristic polynomial of the aircraft can be written as $s^n + a_{n-1}s^{n-1} + \cdots + a_1 s + a_0$. Through analysis, the relationships between the coefficients $a_{n-1}, \ldots, a_1, a_0$ and the degree of the control surface impairment can be derived. Such information provides bounds for the coefficients, from which, an LPV approach can be employed to develop the fault model and synthesize an FTC accordingly.

As different degrees of impairment may occur, for design purposes, the fault model of control surface impairments can be effectively described as an LPV model with the parameter ρ, whose vertices consist of the normal and faulty models. Consequently, the aircraft model with control surface impairments can be described by:

$$\begin{cases} \dot{\mathbf{x}}(t) = \mathbf{A}(\rho)\mathbf{x}(t) + \mathbf{G}(\rho)\omega(t) + \mathbf{B}(\rho)\mathbf{u}(t) \\ \mathbf{y}(t) = \mathbf{C}(\rho)\mathbf{x}(t) \end{cases}, \qquad (3.7)$$

where $\mathbf{x}(t) = \begin{bmatrix} \alpha & \beta & p & q & r \end{bmatrix}^T \in \mathbf{R}^n$ is the state vector of the ADMIRE aircraft, $\omega(t) \in \mathbf{R}^r$ models the external disturbance with bounded energy. $\mathbf{u}(t) =$

$\begin{bmatrix} \delta_{rc} & \delta_{lc} & \delta_{roe} & \delta_{rie} & \delta_{lie} & \delta_{loe} & \delta_r \end{bmatrix}^T \in \mathbf{R}^m$ denotes the control surface deflection bounded by $\mathbf{u}_{\min} \leq \mathbf{u}(t) \leq \mathbf{u}_{\max}$, and $\mathbf{y}(t) = \begin{bmatrix} \alpha & \beta & p \end{bmatrix}^T \in \mathbf{R}^p$ represents the measurement output vector. $\mathbf{A}, \mathbf{B}, \mathbf{C}$ are matrices representing system dynamics, control input, and system output, respectively. $\mathbf{G}$ is a matrix that reflects the distributions of the exogenous disturbances. The linearized aircraft model can be expressed as

$$
\begin{aligned}
(\mathbf{A}(\rho), \mathbf{B}(\rho), \mathbf{C}(\rho), \mathbf{G}(\rho)) &= \sum_{i=1}^{N} \lambda_i (\mathbf{A}_i, \mathbf{B}_i, \mathbf{C}_i, \mathbf{G}_i) \\
&\in Co\{(\mathbf{A}_i, \mathbf{B}_i, \mathbf{C}_i, \mathbf{G}_i): \quad i = 1, \cdots, N\},
\end{aligned} \tag{3.8}
$$

with the convex coordinates $\lambda_i \geq 0$ and $\sum_{i=1}^{N} \lambda_i = 1$. $Co\{\bullet\}$ denotes the convex hull. Furthermore, $(\mathbf{A}_i, \mathbf{B}_i, \mathbf{C}_i, \mathbf{G}_i)$ $(i = 1, \cdots, N)$ are known constant matrices of appropriate dimensions that represent the considered fault models as the vertices. Among these vertices, one corresponds to the normal case, and the remaining vertices correspond to the fault cases. Without loss of generality, $(\mathbf{A}(\rho_1), \mathbf{B}(\rho_1), \mathbf{C}(\rho_1), \mathbf{G}(\rho_1)) = (\mathbf{A}_1, \mathbf{B}_1, \mathbf{C}_1, \mathbf{G}_1)$ is denoted as the nominal aircraft model. The LPV system with bounded parameter $\underline{\rho} \leq \rho \leq \bar{\rho}$ can be represented in a polytopic form when the parameter ρ evolves in a polytopic domain Θ of vertices $[\rho_1, \rho_2, \ldots, \rho_\nu]$ (where the vertices are the extreme values of the parameter ρ). Then, the polytopic LPV system is scheduled through the functions designed as follows:

$$
\Omega = \left\{ \lambda_i \in \Re^N, \lambda_i \geq 0, \sum_{i=1}^{N} \lambda_i = 1 \right\}, \tag{3.9}
$$

where Ω is a convex set, λ_i is the function of varying parameter ρ, and it can be selected using the method proposed in Ref. [84]. In this chapter with the control surface area ρ denoted as the varying parameter, if $N = 2$, $\lambda_1 = (\rho - \underline{\rho}) / (\bar{\rho} - \underline{\rho})$, $\lambda_2 = (\bar{\rho} - \rho) / (\bar{\rho} - \underline{\rho})$.

The control objective is to synthesize the fault-tolerant controllers via the state feedback and the static output feedback to compensate for the control surface impairments.

3.3   Redundancy Analysis and Problem Formulation

A necessary condition for the existence of an FTC is that the system has to possess sufficient actuator redundancies to accommodate the failed control surface. In essence, FTCS design is how to manage existing redundancies in such a way that the integrity of the entire system is maintained. Therefore, the concept of redundancy is introduced in this context front.

3.3.1   Redundancy Analysis

In order to meet the requirements for safety and reliability in the event of failures, the existence of redundancies is the key. Under normal conditions, the advantages of having redundant actuators may not be that obvious; however, they will play an ultimate role in counteracting failures. For clarification, some definitions for redundancy are provided herein. The relationships between the system inputs and outputs, known as functional controllability [85], are utilized to analyze the actuator redundancies.

Definition 3.1. *[85] A system is said to be functionally controllable, if for any given suitable output vector y(defined fort > 0), there exists an input vector u(defined for t > 0) for which this output vector can be achieved from the zero initial condition .*

Remark 3.1. *This definition, in fact, guarantees the independent control of the system outputs by the system inputs. In order to satisfy the functional controllability, the number of independent inputs must be greater than or equal to the number of outputs being controlled.*

Definition 3.2. *[27] For a system described as Eq. (3.7), it is said to have $(m - p)$ degrees of actuator redundancy if the pair (A, b_i) is completely controllable $\forall i\, (1 \le i \le m)$, where b_i is the ith column of B, and the number of independent control inputs is more than that of the system outputs being controlled.*

In a modern aircraft, for instance, several elevons are configured on the wings to provide necessary control functions for pitch and roll motions. However, if malfunctions occur in some elevons, the FTC is able to use the remaining redundant actuators to counteract the failures. Hence, the elevons constitute a set of redundant actuators.

3.3.2   Problem Statement

After the development of a fault aircraft model and the completion of redundancy analysis, the design problem is to find the corresponding state/output feedback FTC such that, for all ρ defined in Chapter 3.2.3,

1) The closed-loop system is stable;

2) The desired output $\mathbf{S}_r \mathbf{y}\,(t)$ tracks the reference signal $\mathbf{r}\,(t)$ with zero steady-state error, that is, $\lim_{t\to\infty} \mathbf{e}\,(t) = 0$, where $\mathbf{e}\,(t) = \mathbf{r}\,(t) - \mathbf{S}_r \mathbf{y}\,(t)$, $\mathbf{S}_r \in \mathbf{R}^{l\times p}$ is a known constant matrix which is used to define specific outputs for tracking control purposes. The reference input can be defined as $\mathbf{r}\,(t) = \sum_{i=1}^{N} \lambda_i \mathbf{r}_i\,(t)$.

The overall structure of the proposed FTCS is shown in Fig. 3.2. A feedback strategy is selected for reconfigurable control system design. Only the pre-designed FTC laws against control surface impairments are investigated in this chapter. It is highlighted in the area enclosed by the dashed lines.

An integral control action is used to eliminate the steady-state tracking error. When the state feedback controller is designed, α, β, p, q, r are used for feedback. In the case of the static output feedback, only the outputs α, β, p are fed back for closed-loop control. In order to obtain a tracking controller with state/output feedback plus the tracking error integral, the augmented tracking system is introduced as

$$
\begin{cases}
\begin{bmatrix} \mathbf{e}(t) \\ \dot{\mathbf{x}}(t) \end{bmatrix} = \begin{bmatrix} \mathbf{0} & -\mathbf{S}_r\mathbf{C}(\rho) \\ \mathbf{0} & \mathbf{A}(\rho) \end{bmatrix} \begin{bmatrix} \int \mathbf{e}(t)\, dt \\ \mathbf{x}(t) \end{bmatrix} + \begin{bmatrix} \mathbf{0} \\ \mathbf{B}(\rho) \end{bmatrix} \mathbf{u}(t) + \begin{bmatrix} \mathbf{I} & \mathbf{0} \\ \mathbf{0} & \mathbf{G}(\rho) \end{bmatrix} \begin{bmatrix} \mathbf{r}(t) \\ \omega(t) \end{bmatrix} \\
\begin{bmatrix} \int \mathbf{e}(t)\, dt \\ \mathbf{y}(t) \end{bmatrix} = \begin{bmatrix} \mathbf{I} & \mathbf{0} \\ \mathbf{0} & \mathbf{C}(\rho) \end{bmatrix} \begin{bmatrix} \int \mathbf{e}(t)\, dt \\ \mathbf{x}(t) \end{bmatrix}
\end{cases}
$$

$$\tag{3.10}$$

Define the augmented state vector $\mathbf{x}_a(t) = \left[\left(\int \mathbf{e}(t)\, dt \right)^T, \mathbf{x}^T(t) \right]^T$, the disturbance vector $\omega_a(t) = \left[\mathbf{r}^T(t), \omega^T(t) \right]^T$, and the augmented measured output vector $\mathbf{y}_a(t) = \left[\left(\int \mathbf{e}(t)\, dt \right)^T, \mathbf{y}^T(t) \right]^T$. Then, Eq. (3.10) can be rewritten as

$$
\begin{cases}
\dot{\mathbf{x}}_a(t) = \mathbf{A}_a(\rho)\mathbf{x}_a(t) + \mathbf{B}_a(\rho)\mathbf{u}_a(t) + \mathbf{G}_a(\rho)\omega_a(t) \\
\mathbf{y}_a(t) = \mathbf{C}_a(\rho)\mathbf{x}_a(t)
\end{cases}, \tag{3.11}
$$

where

$$
\begin{aligned}
\mathbf{A}_a(\rho) &= \begin{bmatrix} \mathbf{0} & -\mathbf{S}_r\mathbf{C}(\rho) \\ \mathbf{0} & \mathbf{A}(\rho) \end{bmatrix} \in \mathbf{R}^{(l+n)\times(l+n)}, \\
\mathbf{B}_a(\rho) &= \begin{bmatrix} \mathbf{0} \\ \mathbf{B}(\rho) \end{bmatrix} \in \mathbf{R}^{(l+n)\times m}, \\
\mathbf{G}_a(\rho) &= \begin{bmatrix} \mathbf{I} & \mathbf{0} \\ \mathbf{0} & \mathbf{G}(\rho) \end{bmatrix} \in \mathbf{R}^{(l+n)\times(l+r)}, \\
\mathbf{C}_a(\rho) &= \begin{bmatrix} \mathbf{I} & \mathbf{0} \\ \mathbf{0} & \mathbf{C}(\rho) \end{bmatrix} \in \mathbf{R}^{(l+p)\times(l+n)}.
\end{aligned}
\tag{3.12}
$$

From Eqs. (3.8) and (3.12), the augmented system can be presented in a polytopic form as follows:

$$
\begin{aligned}
(\mathbf{A}_a(\rho), \mathbf{B}_a(\rho), \mathbf{C}_a(\rho), \mathbf{G}_a(\rho)) &= \sum_{i=1}^{N} \lambda_i (\mathbf{A}_{ai}, \mathbf{B}_{ai}, \mathbf{C}_{ai}, \mathbf{G}_{ai}) \\
&\in Co\left\{ (\mathbf{A}_{ai}, \mathbf{B}_{ai}, \mathbf{C}_{ai}, \mathbf{G}_{ai}): \quad i = 1, \cdots, N \right\},
\end{aligned}
\tag{3.13}
$$

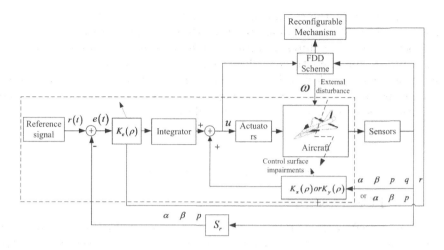

FIGURE 3.2: Overall structure of proposed FTCS.

where

$$\mathbf{A}_{ai} = \begin{bmatrix} \mathbf{0} & -\mathbf{S}_r \mathbf{C}_i \\ \mathbf{0} & \mathbf{A}_i \end{bmatrix} \in \mathbf{R}^{(l+n)\times(l+n)},$$

$$\mathbf{B}_{ai} = \begin{bmatrix} \mathbf{0} \\ \mathbf{B}_i \end{bmatrix} \in \mathbf{R}^{(l+n)\times m},$$

$$\mathbf{G}_{ai} = \begin{bmatrix} \mathbf{I} & \mathbf{0} \\ \mathbf{0} & \mathbf{G}_i \end{bmatrix} \in \mathbf{R}^{(l+n)\times(l+r)}, \tag{3.14}$$

$$\mathbf{C}_{ai} = \begin{bmatrix} \mathbf{I} & \mathbf{0} \\ \mathbf{0} & \mathbf{C}_i \end{bmatrix} \in \mathbf{R}^{(l+p)\times(l+n)}.$$

Choose the regulated output $\mathbf{z}_a(t)$, defined by

$$\mathbf{z}_a(t) = \mathbf{C}_z \mathbf{x}_a(t) + \mathbf{D}_z \mathbf{u}(t). \tag{3.15}$$

Based on the redundancy analysis and the problem formulation, the problem of FTCS design will be addressed next. It should be mentioned that the dynamics of an aircraft depend not only on dynamic pressure, altitude, and airspeed, but also on the available area of control surface. During a normal flight, gain scheduled controllers are used to deal with different aerodynamics due to changes in dynamic pressure, altitude, and airspeed, while flying through various flight regimes. However, these flight regimes related aerodynamic changes should not be confused with those induced by loss (partial loss) of control surfaces. Regardless what regimes the aircraft is flying in, a loss of control surface area has to be accommodated through fault-tolerant flight control systems, which are the main work in this chapter. The loss in control surface area can be estimated by a parameter identification method

[86]. Subsequently, the reconfigurable control strategy can be designed based on the models of the post-fault aircraft (with the estimated surface damage) to compensate for the failures. If a transition in flight regimes has also occurred, the gain scheduled controller will become the base controller for the reconfigurable controller design. For the sake of simplicity, we have not considered the regime changes in the current chapter, because the use of gain scheduled controllers is a common practice in all flight control systems.

3.4 FTCS Design

The aircraft with control surface impairments is modeled as a polytopic LPV system. The scheduling parameter is directly related to the extent of the control surface damage. The advantage of LPV control is that the controller can be designed to maintain the stability of the closed-loop system over a large range of failure conditions. Therefore, the LPV control synthesis method with the pre-defined fault models is used to design an FTC. First, the form of the tracking controller and the closed-loop system is represented. Lemma 3.1 and Theorem 3.1 are then presented for deriving the sufficient conditions for stabilization of the closed-loop system. Subsequently, based on the conclusion of Theorem 3.1, Theorems 3.2 and 3.3 provide two sufficient conditions for synthesizing fault-tolerant controllers via state feedback and static output feedback, respectively.

Consider the augmented system in Eqs. (3.11) and (3.15) with the following state feedback and static output feedback tracking controllers

$$\mathbf{u}(t) = \mathbf{K}(\rho)\mathbf{x}_a(t) = \mathbf{K}_e(\rho) \int \mathbf{e}(t)\, dt + \mathbf{K}_x(\rho)\mathbf{x}(t), \qquad (3.16)$$

$$\mathbf{u}(t) = \mathbf{K}_{SOF}(\rho)\mathbf{y}_a(t) = \mathbf{K}_e(\rho) \int \mathbf{e}(t)\, dt + \mathbf{K}_y(\rho)\mathbf{y}(t), \qquad (3.17)$$

where $\mathbf{K}(\rho) = [\mathbf{K}_e(\rho), \mathbf{K}_x(\rho)] \in \mathbf{R}^{m \times (l+n)}$ and $\mathbf{K}_{SOF}(\rho) = [\mathbf{K}_e(\rho), \mathbf{K}_y(\rho)] \in \mathbf{R}^{m \times (l+p)}$. The closed-loop augmented aircraft equations are given by

$$\begin{aligned} \dot{\mathbf{x}}_a(t) &= \mathbf{A}_{\mathrm{acl}}(\rho)\mathbf{x}_a(t) + \mathbf{G}_a(\rho)\omega_a(t) \\ \mathbf{z}_a(t) &= \mathbf{C}_{\mathrm{acl}}(\rho)\mathbf{x}_a(t) \end{aligned} \qquad (3.18)$$

For the state feedback case, $\mathbf{A}_{acl}(\rho) = \mathbf{A}_a(\rho) + \mathbf{B}_a(\rho)\mathbf{K}(\rho)$, $\mathbf{C}_{acl}(\rho) = \mathbf{C}_z(\rho) + \mathbf{D}_z(\rho)\mathbf{K}(\rho)$, and for the static output feedback case, $\mathbf{A}_{acl}(\rho) = \mathbf{A}_a(\rho) + \mathbf{B}_a(\rho)\mathbf{K}_{SOF}(\rho)\mathbf{C}_a(\rho)$, $\mathbf{C}_{acl}(\rho) = \mathbf{C}_z(\rho) + \mathbf{D}_z(\rho)\mathbf{K}_{SOF}(\rho)\mathbf{C}_a(\rho)$.

Lemma 3.1. *[87] Given a symmetric matrix*

$$\mathbf{H} = \mathbf{H}^T \in \mathbf{R}^{f \times f},$$

and two matrices $\boldsymbol{\Psi}, \boldsymbol{\Phi}$ *of column dimension* f, *there exists a matrix* $\mathbf{M}$ *such that the following LMI holds*

$$\mathbf{H} + \boldsymbol{\Psi}^T \mathbf{M}^T \boldsymbol{\Phi} + \boldsymbol{\Phi}^T \mathbf{M} \boldsymbol{\Psi} < 0. \tag{3.19}$$

$\mathbf{N}_{\boldsymbol{\Psi}}$, $\mathbf{N}_{\boldsymbol{\Phi}}$ *are denoted as arbitrary bases of the null spaces of* $\boldsymbol{\Psi}$ *and* $\boldsymbol{\Phi}$, *respectively. Then (3.19) is solvable if and only if the following projection inequalities are satisfied*

$$\begin{cases} \mathbf{N}_{\boldsymbol{\Psi}}^T \mathbf{H} \mathbf{N}_{\boldsymbol{\Psi}} < 0 \\ \mathbf{N}_{\boldsymbol{\Phi}}^T \mathbf{H} \mathbf{N}_{\boldsymbol{\Phi}} < 0 \end{cases}. \tag{3.20}$$

Theorem 3.1. *Consider the closed-loop system in Eq. (3.18). For a given scalar* $\gamma > 0$ *for all nonzero* $\omega_a(t) \in \mathbf{L}_2[0, \infty)$ *and defining the performance index as* $J(\omega_a) = \int_0^\infty \left(\mathbf{z}_a^T(t)\mathbf{z}_a(t) - \gamma^2 \omega_a^T(t)\omega_a(t) \right) dt < 0$, *if there exists a symmetric positive definite matrix* $\mathbf{P}(\rho) = \mathbf{P}^T(\rho) > 0$ *and any appropriately dimensioned matrix* $\mathbf{V}(\rho)$ *such that the following LMI holds:*

$$\Pi(\rho) = \begin{bmatrix} \begin{array}{c} -\mathbf{A}_{acl}^T(\rho)\mathbf{V}^T(\rho) \\ -\mathbf{V}(\rho)\mathbf{A}_{acl}(\rho) \end{array} & -\mathbf{V}(\rho)\mathbf{G}_a(\rho) & \begin{array}{c} \mathbf{P}(\rho) + \mathbf{V}(\rho) \\ -\mathbf{A}_{acl}^T(\rho)\mathbf{V}^T(\rho) \end{array} & \mathbf{C}_{acl}^T(\rho) \\ * & -\gamma^2\mathbf{I} & -\mathbf{G}_a^T(\rho)\mathbf{V}^T(\rho) & 0 \\ * & * & \left(\mathbf{V}(\rho) + \mathbf{V}^T(\rho)\right) & 0 \\ * & * & * & -\mathbf{I} \end{bmatrix} < 0. \tag{3.21}$$

Then, the closed-loop system in Eq. (3.18) is stable.

Proof. See Appendix C. □

Remark 3.2. *From the proof, it is concluded that the LMI defined in Eq. (3.21) is equivalent to the LMI which is referred from [21] as follows:*

$$\begin{bmatrix} \mathbf{P}(\rho)\mathbf{A}_{acl}(\rho) + \mathbf{A}_{acl}^T(\rho)\mathbf{P}(\rho) & \mathbf{P}(\rho)\mathbf{G}_a(\rho) & \mathbf{C}_{acl}^T(\rho) \\ * & -\gamma^2\mathbf{I} & 0 \\ * & * & -\mathbf{I} \end{bmatrix} < 0, \tag{3.22}$$

where $\mathbf{P}(\rho) = \mathbf{P}^T(\rho) = \sum_{i=1}^N \lambda_i \mathbf{P}_i$ *is a symmetric positive definite matrix. A new form of BRL representation is given by Eq. (3.21). The advantage of the LMI in Eq. (3.21) lies in the fact that by introducing an additional variable* $\mathbf{V}(\rho)$, *the product terms between the Lyapunov matrix* $\mathbf{P}(\rho)$ *and the system matrices* $\mathbf{A}_a(\rho), \mathbf{B}_a(\rho), \mathbf{G}_a(\rho)$ *are eliminated. This decoupling enables the improved BRL to be derived for the polytopic system by using parameter dependent Lyapunov functions, from which Theorems 3.2 and 3.3 can be obtained.*

3.4.1   FTC Design via State Feedback

In the state feedback case, Theorem 3.2 is derived based on Theorem 3.1. The parameter represents the total control surface area. This area is directly related to the amount of control derivatives produced during a flight. The aerodynamic stability and control derivatives under the normal condition can be obtained from the wind tunnel test data at the aircraft design and test stages. In an actual flight, the control derivative will depend on the dynamic pressure and the angle of attack of the aircraft through a set of force coefficients [88] The force coefficients can be estimated by using parameter identification methods in real-time to calculate the actual control derivative generated by the actual control surface area. If there is a loss in the control surface, the calculated control derivative will be different from the expected. This difference can then be used to compute the true value of ρ for use in fault-tolerant flight control systems described in this chapter. The detailed procedures of estimating the force coefficients have been described in Refs. [86] and [88], where a frequency-based online parameter identification technique is proposed in Ref. [86], and a neural network-based method is considered in Ref. [88].

Consider Eqs. (3.11) and (3.15), assuming that the degree of control surface impairment can be estimated. The objective is to determine a state feedback control signal $\mathbf{u}\left(t\right) = \mathbf{K}\left(\rho\right)\mathbf{x}_a\left(t\right)$ so that the closed-loop system described in Eq. (3.18) is stable against the control surface impairments, and the performance criteria $\|\mathbf{T}\left(\rho\right)\|_\infty < \gamma$, where $\mathbf{A}_{acl}\left(\rho\right) = \mathbf{A}_a\left(\rho\right) + \mathbf{B}_a\left(\rho\right)\mathbf{K}\left(\rho\right)$, $\mathbf{C}_{acl}\left(\rho\right) = \mathbf{C}_z\left(\rho\right) + \mathbf{D}_z\left(\rho\right)\mathbf{K}\left(\rho\right)$, $\mathbf{T}\left(\rho\right)$ represents the transfer function of the closed-loop system.

Theorem 3.2. *Consider the polytopic LPV aircraft model described in Eqs. (3.14) and (3.18), if there exist matrices $\mathbf{S}_i, \mathbf{L}_i$, symmetric positive matrices $\mathbf{X}_i > 0$ $(i = 1, \cdots, N)$ and a scalar $\gamma > 0$ such that the following LMIs hold:*

$$\mathbf{\Pi}_{ij} + \mathbf{\Pi}_{ji} < 0, \left(1 \leq i \leq j \leq N\right), \tag{3.23}$$

where

$$\mathbf{\Pi}_{ij} = \begin{bmatrix} -\begin{pmatrix} \mathbf{S}_i\mathbf{A}_{aj}^T + \mathbf{L}_i^T\mathbf{B}_{aj}^T \\ + \mathbf{A}_{aj}\mathbf{S}_i^T + \mathbf{B}_{aj}\mathbf{L}_i \end{pmatrix} & -\mathbf{G}_{aj} & \begin{matrix} \mathbf{X}_j + \mathbf{S}_j - \\ \left(\mathbf{S}_i\mathbf{A}_{aj}^T + \mathbf{L}_i^T\mathbf{B}_{aj}^T\right) \end{matrix} & \mathbf{S}_i\mathbf{C}_z^T + \mathbf{L}_i^T\mathbf{D}_z^T \\ * & -\gamma^2\mathbf{I} & -\mathbf{G}_{aj}^T & 0 \\ * & * & \mathbf{S}_j + \mathbf{S}_j^T & 0 \\ * & * & * & -\mathbf{I} \end{bmatrix}. \tag{3.24}$$

Then, the closed-loop system in Eq. (3.18) is stable over all parameter variations ρ and the infinity norm of the transfer function $\|\mathbf{T}\left(\rho\right)\|_\infty < \gamma$ under the state feedback FTC $\mathbf{u}\left(t\right) = \mathbf{K}\left(\rho\right)\mathbf{x}_a\left(t\right)$ with $\mathbf{K}\left(\rho\right) = \left(\sum_{i=1}^N \lambda_i\mathbf{L}_i\right) \bullet \left(\sum_{i=1}^N \lambda_i\mathbf{S}_i^T\right)^{-1}$.

Proof. See Appendix C.   □

Remark 3.3. *Theorem 3.2 provides a sufficient condition for the existence of FTC based on the state feedback in the presence of control surface impairments. The strategy of determining the scheduling parameter λ_i is given in Chapter 3.2.3. Since each fault scenario is represented as a vertex of the polytope, the FTC obtained based on Theorem 3.2 is able to accommodate the considered aircraft failures. The conservatism of this design approach can be measured by the achievable performance. The philosophy of the proposed FTC is to use different FTC laws to deal with different fault situations. Compared with a single reliable controller for various failures [21, 80, 81, 82], this method has potential to result in less conservative controllers than the reliable controller.*

3.4.2   FTC via Static Output Feedback

Even though the state feedback can provide an elegant solution, the states are not always available due to limitations in measurement devices. In practical control system design, an output feedback is much more attractive. In the static output feedback case, Theorem 3.3 can also be derived based on Theorem 3.1.

Consider Eqs. (3.11) and (3.15), assuming that the degree of control surface impairment can be estimated. The objective is to design the static output feedback controller $\mathbf{u}(t) = \mathbf{K}_{SOF}(\rho)\mathbf{y}_a(t)$ so that the closed-loop system in Eq. (3.18) is stable, and the performance index $\|\mathbf{T}(\rho)\|_\infty < \gamma$, where $\mathbf{A}_{acl}(\rho) = \mathbf{A}_a(\rho) + \mathbf{B}_a(\rho)\mathbf{K}_{SOF}(\rho)\mathbf{C}_a(\rho)$, $\mathbf{C}_{acl}(\rho) = \mathbf{C}_z(\rho) + \mathbf{D}_z(\rho)\mathbf{K}_{SOF}(\rho)\mathbf{C}_a(\rho)$, $\mathbf{T}(\rho)$ is the transfer function of the closed-loop system. It is noted that $\mathbf{C}_a(\rho) = \mathbf{C}_{a1} = \cdots = \mathbf{C}_{aN} = \mathbf{C}_a$ in the ADMIRE aircraft model with the full row rank. Let $\mathbf{T}$ be an invertible matrix, such that $\mathbf{C}_a\mathbf{T} = \begin{bmatrix} \mathbf{I} & \mathbf{0} \end{bmatrix}$.

Theorem 3.3. *Consider the polytopic LPV aircraft model described in Eqs. (3.11) and (3.15), if there exist symmetric matrices $\mathbf{Y}_i, 1 \le i \le N$, matrices $\mathbf{S}_i, \mathbf{L}_i, 1 \le i \le N$ with*

$$\mathbf{S}_i = \begin{bmatrix} \mathbf{S}_{11i} & \mathbf{0} \\ \mathbf{S}_{21i} & \mathbf{S}_{22i} \end{bmatrix}, \mathbf{L}_i = \begin{bmatrix} \mathbf{L}_{1i} & \mathbf{0} \end{bmatrix}, \tag{3.25}$$

and $\gamma > 0$ satisfying the following LMIs:

$$\mathbf{\Pi}_{ij} + \mathbf{\Pi}_{ji} < 0 \, (1 \le i \le j \le N), \tag{3.26}$$

where

$$\mathbf{\Pi}_{ij} = \begin{bmatrix} -(\mathbf{A}_{aj}\mathbf{T}\mathbf{S}_i + \mathbf{B}_{aj}\mathbf{L}_i) & \mathbf{Y}_j + \mathbf{T}\mathbf{S}_j & (\mathbf{C}_z\mathbf{T}\mathbf{S}_i + \mathbf{D}_z\mathbf{L}_i)^T \\ -(\mathbf{A}_{aj}\mathbf{T}\mathbf{S}_i + \mathbf{B}_{aj}\mathbf{L}_i)^T & -\mathbf{G}_{aj} & -(\mathbf{A}_{aj}\mathbf{T}\mathbf{S}_i + \mathbf{B}_{aj}\mathbf{L}_i)^T & \\ * & -\gamma^2\mathbf{I} & -\mathbf{G}_{aj}^T & \mathbf{0} \\ * & * & \mathbf{T}\mathbf{S}_j + (\mathbf{T}\mathbf{S}_j)^T & \mathbf{0} \\ * & * & * & -\mathbf{I} \end{bmatrix}. \tag{3.27}$$

$\mathbf{T} = \begin{bmatrix} \mathbf{C}_a^T(\mathbf{C}_a\mathbf{C}_a^T)^{-1}, \mathbf{C}_a^\perp \end{bmatrix}$ *satisfying* $\mathbf{C}_a\mathbf{T} = \begin{bmatrix} \mathbf{I} & \mathbf{0} \end{bmatrix}$, *and* $\mathbf{T}$ *is nonsingular.*

Then, the aircraft in Eq. (3.18) is stable over all parameter variations ρ. In other words, the designed FTC works with any control surface impairment in the range of the pre-defined models. Subsequently, the FTC gain matrix via static output feedback can be given by:

$$\mathbf{K}_{SOF}(\rho) = \left(\sum_{i=1}^{N} \lambda_i \mathbf{L}_{1i} \right) \left(\sum_{i=1}^{N} \lambda_i \mathbf{S}_{11i} \right)^{-1}. \tag{3.28}$$

Proof. See Appendix C.   □

Remark 3.4. *From Theorem 3.3, a sufficient condition for synthesizing the static output feedback controller is given in terms of Theorem 3.1 and LPV modeling techniques. As a result, the aircraft can be stabilized under the designed FTC via static output feedback in the presence of the considered control surface impairments.*

3.5   Illustrative Examples

As stated in Chapter 3.2.3, the state variables of the ADMIRE benchmark aircraft are selected as AOA, angle of sideslip, roll rate, pitch rate, and yaw rate. The exogenous disturbance is chosen to be a constant vertical gust with a magnitude of 5 m/s. The RC, LC, ROE, RIE, LIE, LOE, and rudder are the control inputs. The tracking selection matrix is chosen as $\mathbf{S}_r = diag\,\{1,\,1,\,1\}$ which means that AOA, sideslip angle, and roll rate are to be tracked. For all examples in this chapter, the command signal for AOA is set at 2 degree magnitude, 11 second duration pulse beginning at 1 sec. The reference signal of sideslip and roll rate is 0. The trimmed values of the ADMIRE aircraft equations are: $M_a = 0.45$, $h = 3000$ m, $V_t = 147.86$ m/s, $\alpha = 3.737\,43\,°$, $\beta = 0$, $T = 0.0752$, $\delta_{rc} = \delta_{lc} = -0.0518\,°$, $\delta_{rie} = -0.036\,18\,°$, $\delta_{roe} = \delta_{lie} = \delta_{loe} = -0.036\,18\,°$, $\delta_r = 0$.

The entire procedure of designing FTC which includes the fault models, redundancy analysis, and FTC synthesis is illustrated. The area of the control surface is used as the parameter, and the scheduling parameter is also a function of that as presented in Chapter 3.2.3.

For the 5th order aircraft system, the corresponding characteristic polynomial is $s^5 + a_4 s^4 + a_3 s^3 + a_2 s^2 + a_1 s + a_0$. The relationship between the extent of damage to the inner elevons and the coefficients a_4, a_3, a_2, a_1, a_0 is illustrated in Fig. 3.3. It can be seen that the coefficients a_4, a_3, a_2, a_1, a_0 vary linearly with the degree of damage (proportional to the area of the inner elevons). Consequently, the maxima and minima on the area of inner elevons can be used as vertices in the polytope to describe the normal and faulty conditions.

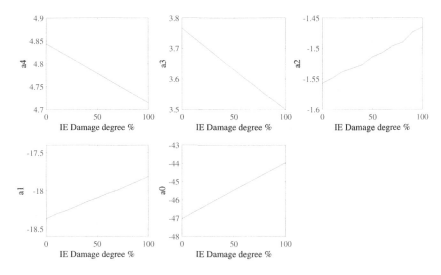

FIGURE 3.3: Coefficients of characteristic polynomials.

The relationship between the area of the inner elevons and the developed fault models is shown in Fig. 3.4. With an increased degree of inner board elevon impairment (decrease of control surface area), the dominant eigenvalues of the fault models change from -3.8744 to -3.8088, which mean the deterioration in stability. The ADMIRE aircraft redundancy is analyzed based on Definitions 3.1 and 3.2 [27, 85]. Since the number of the required outputs is three and the number of independent control inputs is seven, the degree of redundant actuators is four, and the remaining control surfaces are sufficient to counteract the failures.

Several fault scenarios of inner elevon impairment have been considered, but due to the limited space, only the normal case and the worst case corresponding to 100% loss of the inner board elevons are presented in the response figures. In the state feedback example, the detailed performance of the nominal case, inner elevons loss of 50%, inner elevons loss of 75%, and inner elevons loss of 100% are collected in Table 3.1. In Table 3.2, the detailed indices are listed for the static output feedback example. However, in a practical system, the estimation of ρ is not always accurate. Therefore, a sensitivity analysis of the estimation error in ρ is also conducted. The analysis procedure is done by deliberately introducing $\pm 20\%$ error in estimation of ρ, and the performance indices are shown in Tables 3.3 and 3.4 corresponding to the state feedback and static output feedback cases, respectively.

For comparison purposes, a nominal controller that is based on robust control theory [89] and a reliable controller based on the idea in Ref. [21] have been designed. To illustrate the effectiveness of the proposed FTC design approach, the performance of the FTC is compared with that of the robust controller and

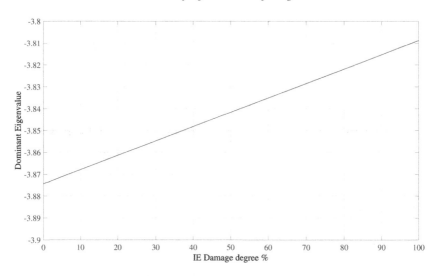

FIGURE 3.4: Relation between elevons damage degree and fault models.

the reliable controller through nonlinear simulations. In the case of the robust controller, the performance is not satisfied when the considered fault occurs. For the reliable controller, only a single controller is designed to deal with all the considered fault situations. The performance under both the normal and faulty cases is worse than that achieved from the proposed FTC. The nominal and fault models of the ADMIRE benchmark aircraft, the robust controllers, the reliable controllers, and the designed fault-tolerant controllers under state feedback and static output feedback cases are given in Appendix D.

3.5.1   Example 1 (State Feedback Case)

Select the following weighting matrices when designing the robust controller, the reliable controller, and the proposed FTC based on LPV techniques:

$$\mathbf{C_z} = [2 \times \mathbf{I}_{6 \times 6}, \mathbf{0}_{6 \times 2}; \mathbf{0}_{7 \times 6}, \mathbf{0}_{7 \times 2}],$$

$$\mathbf{D_z} = [\mathbf{0}_{6 \times 7}; 0.1 \times \mathbf{I}_{7 \times 7}].$$

The design technique for the robust controller is presented in Ref. [86] and the reliable controller is synthesized according to the idea in Ref. [21]. Nonlinear simulation responses of AOA and control surface deflections comparing the FTC and the robust controller are shown in Fig. 3.5. In the interest of space, only the RC deflection responses are displayed. Significant improvement in tracking performance with the FTC is achieved as shown in Figs. 3.5(a) and 3.5(b). The control surface deflections show that the healthy actuator is used to compensate for the effect of inner elevon impairments. Focusing on

TABLE 3.1: AOA performance indices comparison – state feedback.

Controller	Performance	Normal Case	RIE loss 50% LIE loss 50%	RIE loss 75% LIE loss 75%	RIE loss 100% LIE loss 100%
FTC	Overshoot (%)	0.71	3.99	5.68	7.85
	Settling time (s)	0.77	0.74	1.20	1.33
Robust controller	Overshoot (%)	0.03	10.32	21.85	41.08
	Settling time (s)	0.75	1.10	1.83	4.32
Reliable controller	Overshoot (%)	8.37	9.07	9.46	9.86
	Settling time (s)	1.38	1.40	1.40	1.41

the worst case in Fig. 3.5, the tracking performance of AOA is not acceptable for the robust controller. However, the designed FTC performs effectively in this situation. Comparison of the proposed FTC and the reliable controller is provided in Fig. 3.6. It can be seen that the proposed FTC has a superior performance to that of the reliable controller, where zero steady error with a very small transient is obtained. As indicated in Fig. 3.6(c), the remaining control surfaces are stepped up to counteract the negative effects caused by the impairments.

Table 3.1 indicates that the robust controller can still stabilize the system even for the worst case of inner elevon impairments. However, the corresponding performance has deteriorated considerably. The proposed FTC results in an overshoot index of 0.71% and a settling time index of 0.77 s which are not very far from those obtained from the robust controller under the normal case. By contrast, in the worst case, it is indicated that the performance indices corresponding to the proposed FTC with 7.85% overshoot and 1.33 s settling time are significantly better, as compared to those with 41.08% overshoot and 4.32 s settling time from the robust controller.

In the other two fault cases, the FTC has much better tracking performance than that of the robust controller. Moreover, the FTC performs significantly better than the reliable controller with less overshoot and shorter settling time in AOA response. In the worst case, the performances of overshoot and settling time achieved by the proposed FTC are improved by 20.39% (from 9.86% to 7.85%) and 5.67% (from 1.41 s to 1.33 s), respectively. There-

TABLE 3.2: AOA performance indices comparison – static output feedback.

Controller	Performance	Normal Case	RIE 50% LIE 50%	loss loss	RIE 75% LIE 75%	loss loss	RIE 100% LIE 100%	loss loss
FTC	Overshoot (%)	7.11	7.20		7.48		7.52	
	Settling time (s)	4.72	4.75		4.82		5.07	
Robust	Overshoot (%)	0.42	41.07		Oscillatory		Oscillatory	
controller	Settling time (s)	4.23	4.32		Oscillatory		Oscillatory	
Reliable	Overshoot (%)	7.70	7.76		7.81		8.08	
controller	Settling time (s)	4.89	4.98		5.02		5.07	

fore, it is concluded that the proposed FTC is less conservative than the reliable controller.

3.5.2   Example 2 (Static Output Feedback Case)

In this illustrative example, the weight matrices when designing the robust controller, the reliable controller, and the proposed FTC are selected to be:

$$\mathbf{C}_{zz} = diag\,\{2,\,1.2,\,1.2,\,2,\,1,\,1\}\,, \mathbf{D}_{zz} = \mathbf{I}_{7\times 7}, \mathbf{C}_z = \begin{bmatrix} 0.6 \times \mathbf{C}_{zz} & \mathbf{0}_{6\times 2} \\ \mathbf{0}_{7\times 6} & \mathbf{0}_{7\times 2} \end{bmatrix},$$

$$\mathbf{D}_z = \begin{bmatrix} \mathbf{0}_{6\times 7} \\ 0.2 \times \mathbf{D}_{zz} \end{bmatrix}.$$

Only the normal and the worst cases are provided in the response plots, and only RC and LC deflections are shown. The results of a comparison similar to that under the state feedback case are shown in Figs. 3.7 and 3.8, respectively. From the transient responses in Fig. 3.7(b), the FTC performs much better when the considered fault occurs. From the control surface deflection responses, the deflections under the fault case are a little larger than those under the nominal case since the remaining control surfaces are driven to counteract the fault effects. The compared results of the proposed FTC and the reliable controller are illustrated in Fig. 3.8. The comparison curves of the specified scenarios are illustrated to validate the effectiveness of the designed FTC.

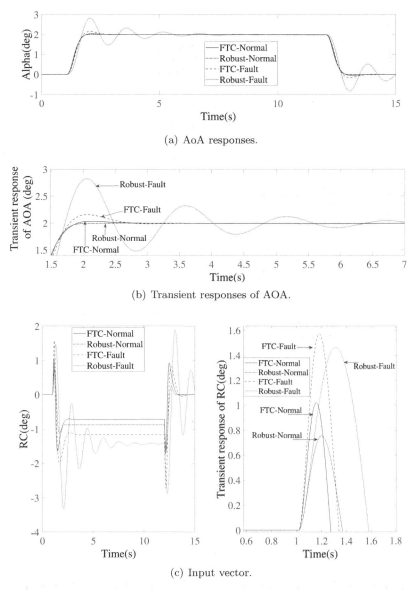

(a) AoA responses.

(b) Transient responses of AOA.

(c) Input vector.

FIGURE 3.5: Comparison results with FTC and robust controller via state feedback in nonlinear simulation.

The detailed performance indices are listed in Table 3.2. In the considered examples, the robust controller cannot stabilize the system when the more serious failures occur. During normal operation, the proposed method results in an overshoot of 7.11% and a settling time of 4.72 s, which are not very far

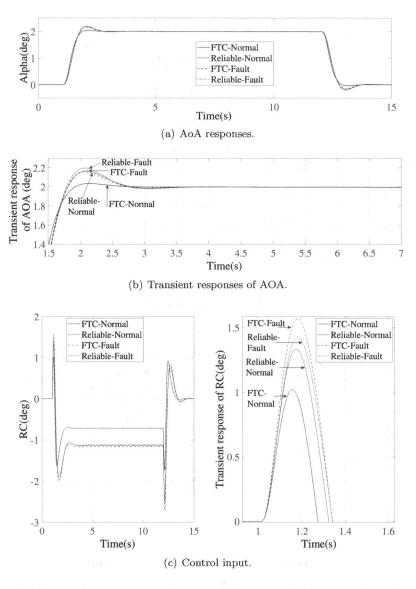

(a) AoA responses.

(b) Transient responses of AOA.

(c) Control input.

FIGURE 3.6: Comparison results with FTC and reliable controller via state feedback in nonlinear simulation.

from the indices obtained by the robust controller. Focus on the performances achieved by the reliable controller and the designed FTC. When compared with the reliable controller, the performance of overshoot with the proposed FTC is improved by 7.66%, 7.22%, 4.23%, and 6.93%. The performance of settling time with FTC is improved by 3.47%, 4.61%, 3.98%, and 0%, respectively. The comparison results confirm that the FTC corresponding to a certain degree of control surface impairments is less conservative than the reliable controller.

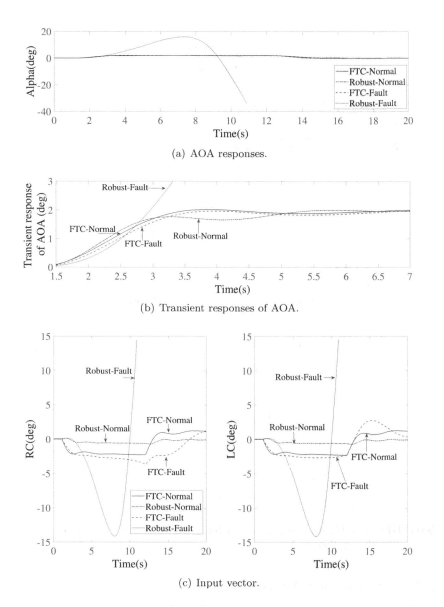

(a) AOA responses.

(b) Transient responses of AOA.

(c) Input vector.

FIGURE 3.7: Comparison results with FTC and robust controller via SOF in nonlinear simulation.

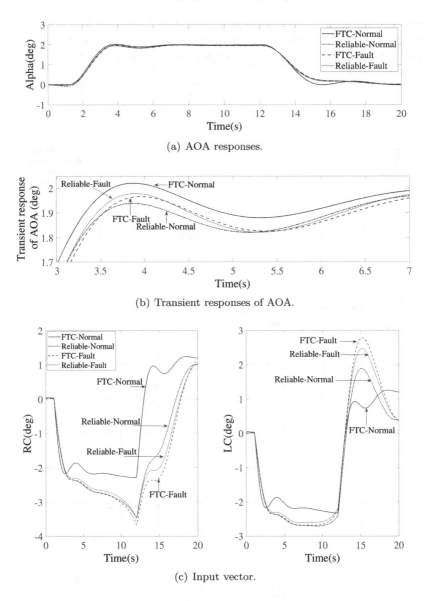

(a) AOA responses.

(b) Transient responses of AOA.

(c) Input vector.

FIGURE 3.8: Comparison results with FTC and reliable controller via SOF in nonlinear simulation.

3.5.3   Sensitivity Analysis

The scheduling parameter is a function of the control surface area. In a practical flight control system, the FTC is scheduled according to the estimate of the remaining control surface area. After the comparison among the pro-

TABLE 3.3: Sensitivity analysis–state feedback.

Controller	Performance	−20% Error in ρ	No Error	+20% Error in ρ
FTC with IE	Overshoot (%)	4.11	3.99	5.22
50% loss	Settling time (s)	0.76	0.74	1.12
FTC with IE	Overshoot (%)	5.81	5.68	7.45
75% loss	Settling time (s)	1.21	1.20	1.29

TABLE 3.4: Sensitivity analysis–static output feedback.

Controller	Performance	−20% Error in ρ	No Error	+20% Error in ρ
FTC with IE	Overshoot (%)	7.83	7.20	10.93
50% loss	Settling time (s)	4.93	4.75	5.31
FTC with IE	Overshoot (%)	7.98	7.48	14.63
75% loss	Settling time (s)	5.22	4.82	6.95

posed FTC, the robust controller, and the reliable controller, the sensitivity analysis of error in estimation of ρ is carried out. In Table 3.3, a $\pm 20\%$ error in estimation is considered for the state feedback case. Through sensitivity analysis of error in the estimated value of ρ, it is validated that the designed FTC can still stabilize the faulty system and achieve an acceptable level of performance when the given degree of estimation error occurs.

Similarly, sensitivity analysis under the static output feedback case is also conducted, and the corresponding indices are indicated in Table 3.4. From the analysis results, the proposed FTC still performs well when the estimation of ρ is not very accurate.

Summarizing the numerical examples of control surface impairment in both the state feedback and the static output feedback cases, it can be seen that the designed FTC can significantly improve the system performance in the event of fault cases, when compared with the robust and the reliable controllers. Furthermore, during nominal operation, the FTC achieves almost the same level of performance as the robust controller. From the sensitivity analysis, the proposed FTC performs satisfactorily in the case of error in estimation of ρ. Therefore, the effectiveness of the proposed approach has been validated by the extensive nonlinear aircraft simulations.

3.6   Conclusions

In this chapter, an FTC design procedure against control surface impairments is investigated. The control surface impairments are modeled as an LPV polytopic model using a parameter which is proportional to the effective control surface. The actuator redundancy of the system has been examined under various control surface impairments. Subsequently, FTC laws are designed using both the state and static output feedbacks. Nonlinear simulations have been carried out using the parameters from the ADMIRE aircraft, and the results have shown that the proposed techniques can indeed maintain the essential properties of the aircraft under various fault conditions. The performance of the FTC has been compared against that of robust controller and reliable controller. It concludes that the conservatism has been considerably reduced.

3.7   Notes

The main contribution of this chapter is to develop the entire procedure of designing an FTCS against control surface impairment failures. The procedure includes developing the aircraft failure models in which the control surface area is considered as a fault parameter, analyzing redundancy under the fault conditions, and designing the FTC based on LPV techniques through the optimization of LMIs to counteract control surface impairments. Compared with robust and reliable controller, the conservatism of the proposed controller has been considerably reduced.

Chapter 4

Multiple Observers Based Anti-Disturbance Control for a Quadrotor UAV

4.1 Introduction

Quadrotor unmanned aerial vehicles (UAVs) have been developed in recent decades [90] and been applied to various fields, including search and rescue, logistics transportation, photography, and agriculture. Multiple disturbances with various characteristics and acting channels can not only affect the control performance of UAV but also deteriorate the UAV safety. How to develop an effective control against multiple disturbances is still a challenging issue. Anti-disturbance control, which is able to compensate/attenuate disturbances and ensure the desired performance, has drawn significant attention.

In Ref. [91], an integrated predictive control $/H_\infty$ control structure is proposed. Based on the solutions of Hamilton-Jacobi-Bellman-Isaacs (HJBI) equations, an H_∞ control scheme is designed such that the L_2 gain from disturbance to attitude is less than or equal to a prescribed value. In Ref. [92], the finite frequency H_∞ control is adopted for stabilization of quadrotors while the PID based H_∞ loop shaping control is designed for the linear translation. A decentralized PID neutral network control system is developed in Ref. [93], where a wind model is explicitly considered. Numerical simulation studies are conducted to show the effectiveness of the control methods presented in Refs. [91, 92, 93]. In Ref. [94], switching model predictive control (MPC) is proposed, and the performance is illustrated by quadrotor UAV flight tests. The test results exemplify that the switching MPC outperforms linear quadratic regulator (LQR) and PID. What's more, an iterative learning based control problem is addressed in Ref. [95], which can only attenuate a periodic disturbance.

On the other hand, disturbance estimation and attenuation are regarded as one of the most popular anti-disturbance strategies. In Ref. [96], sliding mode control (SMC) and sliding mode disturbance observer (SMDO) are developed to estimate the external and internal disturbances on the quadrotor UAV. In Ref. [97], the measured value of accelerometer is used to estimate the disturbance on the quadrotor UAV, whereas the performance is degraded as the quadrotor speeds up. Note that only numerical simulation studies are conducted in the aforementioned references. Flight tests are carried out in Ref. [98], where the influence of time delay, model uncertainty, and external disturbance is considered. A disturbance observer is designed by using an integral filter algorithm in this chapter, however, since the average value of disturbance is used, the real-time estimate precision is affected. To overcome the disturbance caused by wind, a linear disturbance observer is proposed in Ref. [99]. Experimental results are presented to illustrate its anti-disturbance performance during hovering. However, this scheme will be compromised when the quadrotor maneuvers aggressively.

Nonlinear disturbance observer (NDO) is firstly proposed in Ref. [100] to estimate unknown friction in nonlinear robotic manipulators, which is suitable for handling a constant disturbance. With respect to disturbances with partially known information, an advanced version of NDO is presented to estimate harmonic with known frequency [35]. Moreover, it is convenient to combine NDO with other control methods [28, 38, 101], capable of improving the robustness of a system. On the other hand, when the disturbances are totally unknown, the active disturbance rejection control (ADRC) [33, 102] could be a suitable choice to estimate the so-called lumped disturbance. In Ref. [103], a single nonlinear extended state observer (ESO) is employed in the attitude control loop to provide an estimate for the so-called total disturbance and a nonlinear PD controller without disturbance observers is used in the position control loop. The ADRC is also adopted for the quadrotor UAV [34] where several ESOs are applied for disturbances estimation. In addition, an enhanced anti-disturbance control scheme by combining DO and ESO is firstly proposed in [104] to mitigate the multiple disturbances for a spacecraft. By taking into account the different disturbance characteristics, a composite hierarchical anti-disturbance control (CHADC) concept is firstly addressed in Ref. [28] where CHADC is configured as a two-part series which consists of a disturbance observer and a base line controller for the nominal system. As a refined anti-disturbance control approach, CHADC is non-fragile to the disturbances in comparison with other control schemes. Different types of control methodologies can be combined with disturbance observers for different performance requirements.

In this chapter, a multiple observers based anti-disturbance control (MOBADC) scheme is applied for a quadrotor UAV subject to both cable-suspended-payload disturbance and wind disturbance. In the position loop, a composite DO and ESO scheme is proposed to reject the hybrid disturbances. In the attitude loop, an ESO is designed for the quadrotor UAV to mitigate the disturbance mainly caused by wind. Hybrid disturbance estimation is applied in three local-loops, namely DO and ESO embedded local-loops in the position loop, and another ESO embedded local-loop in the attitude loop. Without changing the control gains in the case of disturbances imposed, the adjustment of control law depends on the outputs of DO and ESO. Moreover, extensive flight tests are conducted to demonstrate the anti-disturbance performance of the proposed control scheme. The main contributions are summarized as follows:

1. When comparing to the DO [35] and the ESO [34, 103], the proposed MOBADC can ensure that the estimation error of the payload oscillating disturbance converges to zero asymptotically and the estimation error of a derivative-bounded wind disturbance converges into a bounded set;

2. By resorting to another ESO in the attitude loop, the effect of the external disturbance and model uncertainty on the attitude of the quadrotor UAV can be attenuated effectively;

3. Real-world flight tests of the quadrotor UAV in a hybrid disturbances environment demonstrate the effectiveness of the proposed MOBADC, which improves the mean error and standard deviation (STD) by 76.70% and 71.14%, respectively, compared with the classical PID method.

The remainder of the chapter is organized as follows. Chapter 4.2 presents quadrotor UAV dynamics. Chapter 4.3 includes the detailed design procedure of the proposed control scheme. In Chapter 4.4, experimental results are presented. Chapter 4.5 concludes this chapter.

Notation. *Throughout this chapter, $\mathbb{R}$ represents the set of real numbers. For a matrix $\mathbf{A}$, notations $\|\mathbf{A}\|$ and $\mathbf{A}^{\mathrm{T}}$ denote its Euclidean norm and transpose. s_x and c_x denote $\sin x$ and $\cos x$, respectively.*

4.2 Quadrotor Dynamics with Multiple Disturbances

4.2.1 Quadrotor Dynamic Model

The quadrotor dynamic system is a nonlinear, underactuated, and strong coupled system. The quadrotor UAV has six degrees of freedom and is actuated by four motors. Fig. 4.1 shows the quadrotor UAV flying under both cable-suspended-payload disturbance and wind disturbance.

The configuration of a quadrotor UAV is illustrated in Fig. 4.2. Two coordinate systems are defined in this chaptert, namely, world-fixed frame $\mathcal{W} = [e_1, e_2, e_3]$ and body-fixed frame $\mathcal{B} = [b_1, b_2, b_3]$. In addition, we suppose that the origin of the body-fixed frame is located at the center of the quadrotor's mass. $\boldsymbol{\gamma} = [x, y, z]^{\mathrm{T}}$ denotes the position of the quadrotor in the world-fixed frame and its linear velocity is represented by $\boldsymbol{\nu} = [\dot{x}, \dot{y}, \dot{z}]^{\mathrm{T}} = [u, v, w]^{\mathrm{T}}$. $\boldsymbol{\eta} = [\phi, \theta, \psi]^{\mathrm{T}}$ represents the Euler angular between the world frame and the body frame, and its corresponding angular velocity is denoted by $\boldsymbol{\Omega} = [\dot{\phi}, \dot{\theta}, \dot{\psi}]^{\mathrm{T}}$ in the world-fixed frame. $\boldsymbol{\omega} = [p, q, r]^{\mathrm{T}}$ represents the angular velocity of the quadrotor UAV in the body frame.

FIGURE 4.1: The quadrotor UAV flying under both cable-suspended payload and wind disturbance: the weight of the payload is $500\,\text{g}$, 70% of UAV, and the wind speed is up to $5\,\text{m/s}$.

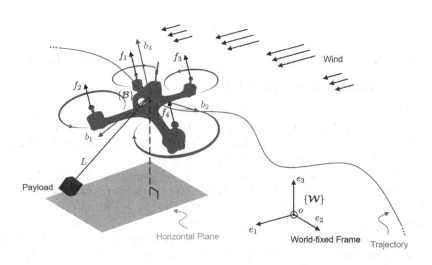

FIGURE 4.2: The quadrotor and the cable-suspended payload in wind environment.

The rotation matrix from the body-fixed frame to the world-fixed frame is given by

$$
\boldsymbol{R} = \begin{bmatrix} c_\psi c_\theta & c_\psi s_\theta s_\phi - s_\psi c_\phi & c_\psi s_\theta c_\phi + s_\psi s_\phi \\ s_\psi c_\theta & s_\psi s_\theta s_\phi + c_\psi c_\phi & s_\psi s_\theta c_\phi - c_\psi s_\phi \\ -s_\theta & s_\phi c_\theta & c_\phi c_\theta \end{bmatrix}. \tag{4.1}
$$

Thrust $f \in \mathbb{R}$ and moment $\boldsymbol{\tau} \in \mathbb{R}^3$, as the control ouputs, can be presented as

$$
\begin{bmatrix} f \\ \tau_\phi \\ \tau_\theta \\ \tau_\psi \end{bmatrix} = \begin{bmatrix} 1 & 1 & 1 & 1 \\ -d_\phi & -d_\phi & d_\phi & d_\phi \\ d_\theta & -d_\theta & d_\theta & -d_\theta \\ c_{\tau f} & -c_{\tau f} & -c_{\tau f} & c_{\tau f} \end{bmatrix} \begin{bmatrix} f_1 \\ f_2 \\ f_3 \\ f_4 \end{bmatrix}, \tag{4.2}
$$

where f_i, $i = 1, 2, 3, 4$ is the thrust force generated by the ith propeller, $\boldsymbol{\tau} = [\tau_\phi, \tau_\theta, \tau_\psi]^\mathrm{T}$ represents the equivalent control moment of the four thrust forces on the quadrotor, d_ϕ is the half of roll motor-to-motor distance, d_θ is the half of pitch motor-to-motor distance, and $c_{\tau f}$ is a fixed constant reflecting the relationship between the thrust force f_i and its corresponding torque.

As can be seen from Fig. 4.2, the quadrotor UAV's translational and rotational motions are driven by adjusting the speed of four motors, while the four lifts (f_1, f_2, f_3, f_4) are proportional to the square of the corresponding rotor speeds. The pitching motion is achieved by decreasing the forces f_2 and f_4 and increasing the forces f_1 and f_3, leading to a forward movement. In a similar manner, the rolling motion is achieved by increasing the forces f_3 and f_4 and decreasing the forces f_1 and f_2, resulting in a lateral movement. Yaw motion is achieved by accelerating the two clockwise turning rotors (f_2, f_3) and decelerating the two counter-clockwise turning rotors (f_1, f_4). The mathematical model can be divided into two portions: translational dynamics and rotational dynamics, respectively.

Based on the Newton's second law, the dynamic of the position loop can be formulated as

$$
m\dot{\boldsymbol{\nu}} = \boldsymbol{F} - mg\boldsymbol{e}_3 + \boldsymbol{d}_f, \tag{4.3a}
$$

$$
\begin{bmatrix} F_x \\ F_y \\ F_z \end{bmatrix} = \frac{f}{m} \begin{bmatrix} c_\psi s_\theta c_\phi + s_\psi s_\phi \\ s_\psi s_\theta c_\phi - s_\phi c_\psi \\ c_\theta c_\phi \end{bmatrix}, \tag{4.3b}
$$

where m is the total mass of the quadrotor, $\boldsymbol{F} = [F_x, F_y, F_z]^\mathrm{T}$ is the equivalent control force of the position loop, g is the gravity acceleration, and $\boldsymbol{d}_f$ stands for the disturbance force vector constituted by environmental disturbance and model uncertainty of the position loop, respectively.

Based on the Lagrange-Euler formalism [91], the dynamic of attitude loop is formulated as

$$
M(\boldsymbol{\eta})\ddot{\boldsymbol{\eta}} + C(\boldsymbol{\eta}, \dot{\boldsymbol{\eta}})\dot{\boldsymbol{\eta}} = \boldsymbol{\tau} + \boldsymbol{d}_\tau, \tag{4.4}
$$

where $M(\boldsymbol{\eta}) \in \mathbb{R}^{3\times3}$ represents the diagonal moment of inertia tensor, and $C(\boldsymbol{\eta}, \dot{\boldsymbol{\eta}}) \in \mathbb{R}^3$ is the centrifugal and Coriolis matrix. $\boldsymbol{d}_\tau$ is the disturbance torque matrix constituted by environmental disturbance and model uncertainty of the attitude loop. The expansions of $M(\boldsymbol{\eta})$ and $C(\boldsymbol{\eta}, \dot{\boldsymbol{\eta}})$ are given by

$$
M(\boldsymbol{\eta}) = \begin{bmatrix} I_{xx} & 0 & -I_{xx}s_\theta \\ 0 & I_{yy}c_\phi^2 + I_{zz}s_\phi^2 & (I_{yy} - I_{zz})c_\phi s_\phi c_\theta \\ -I_{xx}s_\theta & (I_{yy} - I_{zz})c_\phi s_\phi c_\theta & I_{xx}s_\theta^2 + (I_{yy}s_\phi^2 + I_{zz}c_\phi^2)c_\theta^2 \end{bmatrix}
$$

and

$$
C(\boldsymbol{\eta}, \dot{\boldsymbol{\eta}}) = \begin{bmatrix} c_{11} & c_{12} & c_{13} \\ c_{21} & c_{22} & c_{23} \\ c_{31} & c_{32} & c_{33} \end{bmatrix},
$$

where

$$
c_{11} = 0,
$$
$$
c_{12} = (I_{yy} - I_{zz})(\dot{\theta}c_\phi s_\phi + \dot{\psi}s_\phi^2 c_\theta) - (I_{xx} + I_{yy}c_\phi^2 - I_{zz}c_\phi^2)\dot{\psi}c_\theta,
$$
$$
c_{13} = (I_{zz} - I_{yy})\dot{\psi}c_\phi s_\phi c_\theta^2,
$$
$$
c_{21} = (I_{zz} - I_{yy})(\dot{\theta}c_\phi s_\phi + \dot{\psi}s_\phi^2 c_\theta) + (I_{xx} + I_{yy}c_\phi^2 - I_{zz}c_\phi^2)\dot{\psi}c_\theta,
$$
$$
c_{22} = (I_{zz} - I_{yy})\dot{\phi}c_\phi s_\phi,
$$
$$
c_{23} = (-I_{xx} + I_{yy}s_\phi^2 + I_{zz}c_\phi^2)\dot{\psi}s_\theta c_\theta,
$$
$$
c_{31} = (I_{yy} - I_{zz})\dot{\psi}c_\theta^2 s_\phi c_\phi - I_{xx}\dot{\theta}c_\theta,
$$
$$
c_{32} = (I_{zz} - I_{yy})(\dot{\theta}c_\phi s_\phi s_\theta + \dot{\phi}s_\phi^2 c_\theta - \dot{\phi}c_\phi^2 c_\theta),
$$
$$
\quad + (I_{xx} - I_{yy}s_\phi^2 - I_{zz}c_\phi^2)\dot{\psi}s_\theta c_\theta,
$$
$$
c_{33} = (I_{yy} - I_{zz})\dot{\phi}c_\phi s_\phi c_\theta^2 + (I_{xx} - I_{yy}s_\phi^2 - I_{zz}c_\phi^2)\dot{\theta}c_\theta s_\theta.
$$

4.2.2　The Analysis of Disturbances

As the quadrotor UAV with a cable-suspended payload operating in a wind field, there are multifarious disturbances acting on the quadrotor such as the payload, wind, rotor drag, and model uncertainty [37, 105]. According to the characteristics, the disturbances concerned can be divided into two types: $\boldsymbol{d}_m$ represented by an exogenous system (e.g., periodic payload disturbance) and $\boldsymbol{d}_l$ described as a derivative bounded variable (e.g., wind disturbance and model uncertainty).

In most situations, the suspended payload can be regarded as a disturbance with partially known information. For example, when the payload swings or the quadrotor has a periodic trajectory, the disturbance can be seen as a kind of $\boldsymbol{d}_m$. Thus, the dynamics can be described as [35],

$$
\begin{cases} \dot{\boldsymbol{\xi}} = \boldsymbol{A}\boldsymbol{\xi} \\ \boldsymbol{d}_m = \boldsymbol{B}\boldsymbol{\xi} \end{cases}, \tag{4.5}
$$

where $\boldsymbol{A} = \begin{bmatrix} \boldsymbol{A}_x & \boldsymbol{0}_{2\times2} & \boldsymbol{0}_{2\times2} \\ \boldsymbol{0}_{2\times2} & \boldsymbol{A}_y & \boldsymbol{0}_{2\times2} \\ \boldsymbol{0}_{2\times2} & \boldsymbol{0}_{2\times2} & \boldsymbol{A}_z \end{bmatrix}$, $\boldsymbol{A}_i = \begin{bmatrix} 0 & \sigma_i \\ -\sigma_i & 0 \end{bmatrix}_{i=x,y,z}$, $\boldsymbol{B} =$

$\begin{bmatrix} 1 & 0 & 0 & 0 & 0 & 0 \\ 0 & 0 & 1 & 0 & 0 & 0 \\ 0 & 0 & 0 & 0 & 1 & 0 \end{bmatrix}$, and σ_i is the frequency of the periodic disturbance

along i-axis.

Suppose that the lashing point between the quadrotor and the cable is coincident with the center of gravity of the quadrotor, then the periodic disturbance $\boldsymbol{d}_m$ caused by the payload will directly affect the position of the quadrotor UAV. Thus $\boldsymbol{d}_m$ can be classified as part of $\boldsymbol{d}_f$ given in Eq. (4.3a), denoted as $\boldsymbol{d}_{mf}$.

On the other hand, without loss of generality, suppose that the wind disturbance suffered by the quadrotor UAV has a bounded variation and is regarded as $\boldsymbol{d}_l$ in this chapter. The upper bound of $\dot{\boldsymbol{d}}_l$ is denoted by a positive scalar $\bar{d}_l$, i.e., $\|\dot{\boldsymbol{d}}_l\| \leq \bar{d}_l$.

In nature, it is difficult to accurately predict wind disturbance, especially its direction and speed. For a quadrotor UAV, if the wind disturbance invades laterally, it mainly affects the UAV's position. Its attitude will be influenced greatly when the wind blows upwards or downwards and is not precisely aligned with the center of mass. In most cases, the effect of wind disturbance on the quadrotor UAV is a mixture of the preceding two scenarios and its variation can be considered bounded. Thus, corresponding to the force and torque disturbances in Eqs. (4.3a) and (4.4), $\boldsymbol{d}_l$ herein is divided into two types: force disturbance part $\boldsymbol{d}_{lf}$ and torque disturbance part $\boldsymbol{d}_{l\tau}$. The following assumption about wind disturbance is presented.

Assumption 4.1. *The force and torque portions of wind disturbance are assumed to have bounded variation and there exist two constants $\bar{d}_{lf}$ and $\bar{d}_{l\tau}$ such that $\|\dot{\boldsymbol{d}}_{lf}\| \leq \bar{d}_{lf}$ and $\|\dot{\boldsymbol{d}}_{l\tau}\| \leq \bar{d}_{l\tau}$.*

Note that it requires that the changing rate cannot be too large, otherwise the disturbance is difficult to be observed.

To sum up, the translational dynamic Eq. (4.3a) and rotational dynamic Eq. (4.4) of a quadrotor UAV under both cable-suspended-payload disturbance and wind disturbance can be rewritten as

$$\begin{cases} m\dot{\boldsymbol{\nu}} = \boldsymbol{F} - mg\boldsymbol{e}_3 + \boldsymbol{d}_{mf} + \boldsymbol{d}_{lf} \\ M(\boldsymbol{\eta})\ddot{\boldsymbol{\eta}} + C(\boldsymbol{\eta}, \dot{\boldsymbol{\eta}})\dot{\boldsymbol{\eta}} = \boldsymbol{\tau} + \boldsymbol{d}_{l\tau} \end{cases}, \tag{4.6}$$

where $\boldsymbol{d}_f = \boldsymbol{d}_{mf} + \boldsymbol{d}_{lf}$ and $\boldsymbol{d}_\tau = \boldsymbol{d}_{l\tau}$.

With the translational and rotational dynamics of a quadrotor UAV in the presence of the disturbances being modelled, a MDOBAC scheme is presented to achieve high-precision control performance next.

4.3   Design of Multiple Observers Based Anti-Disturbance Control

In order to mitigate the influence of periodic disturbance and wind disturbance, the MOBADC is composed of a DO and an ESO in the position loop, and only ESO based nonlinear control in the attitude loop. The proposed control scheme and data flows can be seen in Fig. 4.3.

4.3.1   Control for Translational Dynamics

Firstly, the errors of position and linear velocity are defined as

$$e_\gamma = \gamma_d - \gamma, e_\nu = \nu_d - \nu, \tag{4.7}$$

where γ_d and ν_d are the desired trajectory and velocity of the quadrotor UAV, and γ and ν are the current position and velocity.

The controller of the position loop is designed as

$$\begin{cases} a_d = K_\gamma e_\gamma + K_\nu e_\nu + g e_3 + \ddot{\gamma}_d \\ F = m a_d - \hat{d}_{mf} - \hat{d}_{lf} \end{cases}, \tag{4.8}$$

where a_d is the desired control acceleration in the position loop, K_γ and K_ν are positive definite gain matrices, $\ddot{\gamma}_d$ is the desired trajectory acceleration. Note that the estimation values $\hat{d}_{mf}$ and $\hat{d}_{lf}$ will be given below.

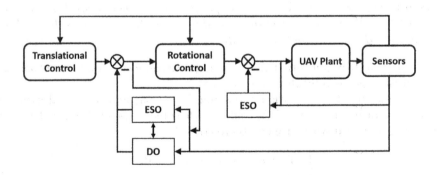

FIGURE 4.3: MOBADC scheme and relevant data flows.

4.3.1.1 DO Design

The DO is designed to estimate the periodic disturbance $\boldsymbol{d}_m$. Based on Eq. (4.3a) and inspired by Refs. [104, 106], the DO is designed as

$$
\begin{cases}
\dot{z} = (A - l(\gamma, \nu)GB)\, z + Ap(\gamma, \nu) - \\
\qquad l(\gamma, \nu) \left(GBp(\gamma, \nu) + f(\nu) + GF + G\hat{d}_{lf} \right) \\
\hat{\xi} = z + p(\gamma, \nu) \\
\hat{d}_{mf} = B\hat{\xi}
\end{cases}, \tag{4.9}
$$

where $\hat{\xi}$ and $\hat{d}_{mf}$ represent the estimates of ξ and $\boldsymbol{d}_{mf}$, z is the auxiliary state variable of the observer, $\hat{d}_{lf}$ denotes the estimate of $\boldsymbol{d}_{lf}$ which can be obtained by ESO, $p(\gamma, \nu)$ is an auxiliary function to be designed, and $l(\gamma, \nu)$ is the gain function of the observer determined by

$$
l(\gamma, \nu) = \left(\frac{\partial p(\gamma, \nu)}{\partial \gamma}, \frac{\partial p(\gamma, \nu)}{\partial \nu} \right), \tag{4.10}
$$

and $\boldsymbol{G} = \frac{1}{m}[\boldsymbol{0}_{3\times3}, \boldsymbol{I}_{3\times3}]^{\mathrm{T}}$, $f(\nu) = [\nu^{\mathrm{T}}, 0, 0, -g]^{\mathrm{T}}$.

4.3.1.2 ESO Design

In order to facilitate the design process of ESO, we first denote $\boldsymbol{x}_{p1} = \gamma$, $\boldsymbol{x}_{p2} = \nu$, and $\boldsymbol{x}_{p3} = \boldsymbol{d}_{lf}$, respectively. Subsequently, the translational dynamic in Eq. (4.3a) can be transformed to a general form [33] as

$$
\begin{cases}
\dot{\boldsymbol{x}}_{p1} = \boldsymbol{x}_{p2} \\
\dot{\boldsymbol{x}}_{p2} = \dfrac{1}{m}(\boldsymbol{F} - mg\boldsymbol{e}_3 + \boldsymbol{x}_{p3} + \boldsymbol{d}_{mf}) \\
\dot{\boldsymbol{x}}_{p3} = \dot{\boldsymbol{d}}_{lf}
\end{cases}, \tag{4.11}
$$

The ESO serving in the position loop is designed to estimate the lumped disturbance $\boldsymbol{d}_{lf}$.

$$
\begin{cases}
\dot{z}_{p1} = z_{p2} + K_{p1}e_{p1} \\
\dot{z}_{p2} = \dfrac{1}{m}(\boldsymbol{F} - mg\boldsymbol{e}_3 + z_{p3} + \hat{d}_{mf}) + K_{p2}e_{p1} \\
\dot{z}_{p3} = K_{p3}e_{p1}
\end{cases}, \tag{4.12}
$$

where z_{p1}, z_{p2}, and z_{p3} are the estimates of $\boldsymbol{x}_{p1}$, $\boldsymbol{x}_{p2}$, and $\boldsymbol{x}_{p3}$, $\boldsymbol{K}_{p1}$, $\boldsymbol{K}_{p2}$, and $\boldsymbol{K}_{p3} \in \mathbb{R}^{3\times3}$ are the observer gains of the ESO in the position loop and $e_{p1} = \boldsymbol{x}_{p1} - z_{p1}$. For the sake of simplification, define the stacked gain matrix $\boldsymbol{K}_p = [\boldsymbol{K}_{p1}, \boldsymbol{K}_{p2}, \boldsymbol{K}_{p3}]$.

The gains of control ($\boldsymbol{K}_\gamma$, $\boldsymbol{K}_\nu$) and the observers ($\boldsymbol{K}_p$, $l(\gamma, \nu)$) can be chosen based on Theorem 4.1.

Remark 4.1. *The DO and ESO in this part are designed to respectively tackle the payload oscillating disturbance and disturbances with bounded derivative value that occur in the position loop. Their respective disturbance estimates are utilized by each other which can be seen from Eqs. (4.9) and (4.12).*

Remark 4.2. *The payload oscillating disturbance is a kind of periodic disturbance and the disturbance like wind is a kind of disturbance with bounded derivative value. Since there is a great difference between these kinds of disturbances, the proposed DO and ESO can effectively distinguish and estimate their respective disturbances.*

4.3.2   Control for Rotational Dynamics

According to Eq. (4.3b), the desired input for the attitude loop is driven by

$$\begin{cases} f = \dfrac{F_z}{c_\theta c_\phi} \\[2mm] \theta_d = \arctan(\dfrac{F_x c_\psi + F_y s_\psi}{F_z}) \\[2mm] \phi_d = \arctan(c_\theta \dfrac{F_x s_\psi - F_y c_\psi}{F_z}) \end{cases} , \qquad (4.13)$$

where θ_d and ϕ_d are the desired pitch and roll angles of the quadrotor, and ψ_d is the desired yaw angle that is set to zero in the chapter.

Define the attitude tracking error as

$$e_\eta = \eta_d - \eta, \qquad (4.14)$$

where $\eta_d = [\phi_d, \theta_d, \psi_d]^{\mathrm{T}}$ and η is the current Euler angular.

Hence, the controller for rotational dynamics is designed as

$$\tau_d = K_\eta e_\eta - K_\omega \omega - \hat{d}_{l\tau}, \qquad (4.15)$$

where K_η and K_ω are positive definite gain matrices, and $\hat{d}_{l\tau}$ is the lumped disturbance estimate, respectively.

Remark 4.3. *The desired attitude velocity ω_d is set to zero, which is applicable to the quadrotor UAV operating within a small angle.*

Denote $x_{a1} = \eta$, $x_{a2} = \Omega$, and $x_{a3} = d_{l\tau}$, the rotational dynamic in Eq. (4.4) can be thereby rewritten as

$$\begin{cases} \dot{x}_{a1} = x_{a2} \\ \dot{x}_{a2} = M^{-1}(\tau - C\dot{\eta} + x_{a3}) \\ \dot{x}_{a3} = \dot{d}_{l\tau} \end{cases} . \qquad (4.16)$$

The ESO in the attitude loop is developed to estimate the lumped disturbance $d_{l\tau}$.

$$\begin{cases} \dot{z}_{a1} = z_{a2} + K_{a1}e_{a1} \\ \dot{z}_{a2} = \hat{M}^{-1}(\tau - \hat{C}z_{a2} + z_{a3}) + K_{a2}e_{a1} \\ \dot{z}_{a3} = K_{a3}e_{a1} \end{cases}, \tag{4.17}$$

where z_{a1}, z_{a2}, and z_{a3} are the estimates of x_{a1}, x_{a2}, and x_{a3}, K_{a1}, K_{a2}, and $K_{a3} \in \mathbb{R}^{3 \times 3}$ are the observer gains of ESO in the attitude loop and $e_{a1} = x_{a1} - z_{a1}$. $\hat{M}$ and $\hat{C}$ are the estimates of M and C with respect to z_{a1}, z_{a2}, and z_{a3}. For the sake of simplification, the stacked gain is denoted as $K_a = [K_{a1}, K_{a2}, K_{a3}]$, which can be chosen based on Theorem 4.2.

Remark 4.4. *The ESO in the attitude loop intends to attenuate the external disturbance and model uncertainty fed into the attitude control of the quadrotor UAV. Hence, the roles that the ESO herein, and the DO and ESO in the position loop are separated.*

Remark 4.5. *The presented MOBADC scheme can be seen as a refined anti-disturbance control. It is easy to tailor such that different types of control methodologies can be integrated with various disturbance observers. The developed scheme can handle the complicated case that multiple disturbances exist in different channels. Such an anti-disturbance control can also be applied to the environment where only one kind of disturbance is present. In such circumstance, the estimate of the other disturbance is negligibly small and has little effect on the control input.*

Remark 4.6. *The developed MOBADC scheme intends to reject multiple disturbances exposed on a UAV. It should be emphasized that either ESO or DO can be chosen according to the disturbance characteristics. In comparison of the studies [34, 35], multiple disturbances can be estimated and thereby rejected within the presented scheme. In contrast to the work [103], multiple observers are exploited in the developed scheme in terms of the explicit disturbance analysis and characteristics. Hence, multiple disturbances can be handled in a delicacy manner. From the aspect of anti-disturbance performance, the conservatism can be substantially reduced especially in the presence of multiple disturbances.*

4.3.3 Stability Analysis

4.3.3.1 Position Loop

Define the estimation errors of DO and ESO as

$$e_\xi = \xi - \hat{\xi}, \tag{4.18a}$$

$$e_p = x_p - z_p, \tag{4.18b}$$

where $x_p = [x_{p1}^{\mathrm{T}}, x_{p2}^{\mathrm{T}}, x_{p3}^{\mathrm{T}}]^{\mathrm{T}}$ and $z_p = [z_{p1}^{\mathrm{T}}, z_{p2}^{\mathrm{T}}, z_{p3}^{\mathrm{T}}]^{\mathrm{T}}$. Next, the stability result is given.

In real applications, for convenience, the auxiliary function $p(\gamma, \nu)$ can be chosen as a linear function such that $l(\gamma, \nu)$ will be a constant matrix and rewritten as l.

Theorem 4.1. *The position control system with the proposed composited DO has a bounded error if the disturbance $\dot{d}_{lf}$ is bounded and there exist the controller gains K_γ, K_ν and the observer gains K_p, l such that the eigenvalues of Ξ have negative real parts, where Ξ is given by*

$$\Xi = \begin{bmatrix} Y_{15 \times 15} & 0_{6 \times 15} \\ 0_{15 \times 15} & I_{3 \times 3} \quad 0_{3 \times 3} \\ -\dfrac{1}{m} \begin{bmatrix} B & E \end{bmatrix} & -K_\gamma \quad -K_\nu \end{bmatrix}, \tag{4.19}$$

with $E = [0_{3 \times 3}, 0_{3 \times 3}, I_{3 \times 3}]$, $Y_{15 \times 15} = \begin{bmatrix} (A - lGB)_{6 \times 6} & (-lGE)_{6 \times 9} \\ 0_{3 \times 6} \\ \dfrac{1}{m} I_{3 \times 3} B & S_{9 \times 9} \\ 0_{3 \times 6} \end{bmatrix}$ *and*

$$S_{9 \times 9} = \begin{bmatrix} -K_{p1} & I_{3 \times 3} & 0_{3 \times 3} \\ -K_{p2} & 0_{3 \times 3} & \dfrac{1}{m} I_{3 \times 3} \\ -K_{p3} & 0_{3 \times 3} & 0_{3 \times 3} \end{bmatrix}.$$

Proof. By taking into account Eqs. (4.3a), (4.9), and (4.10), the differential equation of Eq. (4.18a) is given by

$$\begin{aligned} \dot{e}_\xi &= \dot{\xi} - \dot{\hat{\xi}} \\ &\overset{(a)}{=} A\xi - \left(\dot{z} + l \left[\dot{\gamma}^{\mathrm{T}}, \dot{\nu}^{\mathrm{T}} \right]^{\mathrm{T}} \right) \\ &= A\xi - A\hat{\xi} + lGB\hat{\xi} + lf + lGF + lG\hat{d}_{lf} - l \left[\dot{\gamma}^{\mathrm{T}}, \dot{\nu}^{\mathrm{T}} \right]^{\mathrm{T}} \\ &\overset{(b)}{=} Ae_\xi + lGB\hat{\xi} - lGd_{lf} - lGd_{mf} + lG\hat{d}_{lf} \\ &\overset{(c)}{=} (A - lGB)e_\xi - lGEe_p, \end{aligned} \tag{4.20}$$

where $f := f(\nu)$, and (a), (b), and (c) follow from Eqs. (4.9), (4.6), and (4.5), respectively.

According to Eqs. (4.11) and (4.12), the differential equation of Eq. (4.18b) is given by

$$\dot{e}_p = \dot{x}_p - \dot{z}_p \tag{4.21}$$

$$= S_{9 \times 9} e_p + \begin{bmatrix} 0_{3 \times 3} \\ \dfrac{1}{m} I_{3 \times 3} \\ 0_{3 \times 3} \end{bmatrix} Be_\xi + \begin{bmatrix} 0_{3 \times 3} \\ 0_{3 \times 3} \\ I_{3 \times 3} \end{bmatrix} \dot{d}_{lf}. \tag{4.22}$$

By combining Eqs. (4.20) and (4.22), the dynamics of the estimation errors of the position loop can be presented as

$$
\begin{bmatrix} \dot{e}_\xi \\ \dot{e}_p \end{bmatrix} = Y_{15\times15} \begin{bmatrix} e_\xi \\ e_p \end{bmatrix} + \begin{bmatrix} \mathbf{0}_{12\times3} \\ I_{3\times3} \end{bmatrix} \dot{d}_{lf}. \tag{4.23}
$$

Substitute Eq. (4.8) into Eq. (4.3a), and we can get

$$
m\dot{e}_\nu + mK_\nu e_\nu + mK_\gamma e_\gamma + Be_\xi + Ee_p = 0. \tag{4.24}
$$

According to Eq. (4.23), together with Eq. (4.24), the error dynamics of the entire position system is obtained as

$$
\dot{e} = \begin{bmatrix} \dot{e}_\xi \\ \dot{e}_p \\ \dot{e}_\gamma \\ \dot{e}_\nu \end{bmatrix} = \Xi \begin{bmatrix} e_\xi \\ e_p \\ e_\gamma \\ e_\nu \end{bmatrix} + \begin{bmatrix} \mathbf{0}_{12\times3} \\ I_{3\times3} \\ \mathbf{0}_{6\times3} \end{bmatrix} \dot{d}_{lf}. \tag{4.25}
$$

Apparently, based on Assumption 4.1, if the position controller gains and observer gains are selected appropriately to make the eigenvalues of Ξ have negative real parts, the composited DO has a bounded error [102] and the proof is completed. $\qquad\square$

4.3.3.2 Attitude Loop

Define the estimation errors of ESO in the attitude loop as

$$
e_a = x_a - z_a, \tag{4.26}
$$

where $x_a = \begin{bmatrix} x_{a1}^T, x_{a2}^T, x_{a3}^T \end{bmatrix}^T$ and $z_a = \begin{bmatrix} z_{a1}^T, z_{a2}^T, z_{a3}^T \end{bmatrix}^T$. Before moving on, it is necessary to make the following assumption.

Assumption 4.2. *The quadrotor UAV flies at a small angle and its attitude dynamics can be linearized such that $M(\eta)$ and $C(\eta, \dot{\eta})$ will become constant matrices M_0 and C_0.*

Now, we are in a position to give the following result.

Theorem 4.2. *Under the proposed control scheme in Eq. (4.15) and the designed ESO in Eq. (4.17), the attitude control system at a small angle is asymptotically stable.*

Proof. Based on the small-angle approximation, a linearized attitude control system is obtained. According to the so-called separation principle, design of the state observer can be separated from the controller design of the linear system under certain condition [107]. To proceed, two conditions should be satisfied: (1) the attitude loop using the control scheme in Eq. (4.15) is asymptotically stable in the absence of disturbances; (2) the ESO is stable with appropriately chosen observer gain K_a.

Obviously, in the absence of disturbances, the attitude controller Eq. (4.15) will be reduced to a proportional-differential controller and the condition (1) can be satisfied.

According to Assumption 4.2 and similar to the derivation of Eq. (4.18b), the dynamics of the estimation errors in the attitude loop can be obtained by

$$\dot{e}_a = \dot{x}_a - \dot{z}_a = W_{9\times9}e_a + \begin{bmatrix} \mathbf{0}_{3\times3} \\ \mathbf{0}_{3\times3} \\ \mathbf{I}_{3\times3} \end{bmatrix} \dot{d}_{l\tau}, \qquad (4.27)$$

where $W_{9\times9} = \begin{bmatrix} -\mathbf{K}_{a1} & \mathbf{I}_{3\times3} & \mathbf{I}_{3\times3} \\ -\mathbf{K}_{a2} & -\mathbf{M}_0^{-1}\mathbf{C}_0 & \mathbf{M}_0^{-1} \\ -\mathbf{K}_{a3} & \mathbf{0}_{3\times3} & \mathbf{0}_{3\times3} \end{bmatrix}$.

According to Assumption 4.1, $\mathbf{K}_a$ can be appropriately selected such that the eigenvalues of $W_{9\times9}$ have negative real parts, then the ESO in the attitude loop is stable [102] and the condition (2) is satisfied. In addition, Eq. (4.27) shows that the convergence of the observer does not depend on the variation of the state, hence the above theorem can be proved [107].   □

Remark 4.7. *In real applications, the basic control gains ($\mathbf{K}_\eta$, $\mathbf{K}_\omega$, $\mathbf{K}_\gamma$ and $\mathbf{K}_\nu$) are determined first from the inner loop to outer loop. Then, according to Theorems 4.1 and 4.2, the gains of DO and ESOs are determined respectively.*

Remark 4.8. *Although the stability analysis of the attitude loop is based on the approximate linearization, the proposed ESO based controller still works well for its original model Eq. (4.4) without linearization as demonstrated in the real flight experiments.*

4.4   Flight Experimental Results

Flight tests are carried out to demonstrate the effectiveness of the proposed MOBADC system. The flying arena and system configuration are presented first. Then, four groups of flight tests are carried out as can be observed from in Fig. 4.4. The corresponding performance is evaluated in terms of average absolute position error and standard deviation in comparison with the desired trajectory. The ground truth is provided by OptiTrack system[®][1] (a motion capture system with milimeter-level positioning accuracy).

---

[1]https://optitrack.com/products/flex-13/

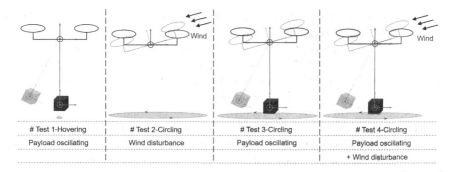

FIGURE 4.4: Four groups of flight tests with different motion scenarios and disturbance types.

4.4.1 Flying Arena and System Configuration

The flying arena (Fig. 4.1) consists of a global sensing (the motion capture system), a base station module with local server, a wireless command interface, off-board estimation and control modules, and on-board sensing, actuation, estimation and control (conducted on QDrone® platform[2]). The connections between each component are presented in Fig. 4.5.

The Intel® Aero Compute Board is utilized as the on-board MCU, which is a purpose-built, UAV developer kit powered by a quad-core Intel® Atom™ processor. The size of a standard playing card features abundant storage capabilities, 802.11ac Wi-Fi, industry standard interfaces, and reconfigurable I/O to facilitate connecting to a broad variety of quadrotor UAV hardware

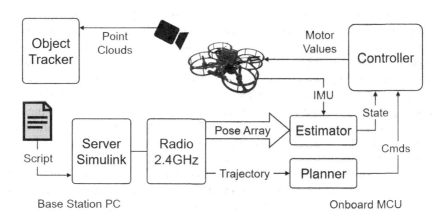

FIGURE 4.5: Diagram of major system components.

---

[2]https://www.quanser.com/products/qdrone/

subsystems. Embedded BM160 IMU, BMM150 magnetometer, and MS5611 barometer sensors support on-board sensing and estimation.

The MCU on quadcopter UAV actively sends on-board sensing data to the base station for high-level mission planning and receives mission commands through a Wi-Fi transceiver module. All of the data collected and estimates obtained by quadrotor will be calibrated and filtered first. Then, the corrected information together with the received mission commands goes through the trajectory generation module. Next, the desired trajectory will be sent to the position control. The outputs will be transformed into attitude command and further into pulse-width modulating (PWM) signals to drive the quadrotor UAV. The above setup and configuration serve for the following flight experiments followed by performances analysis.

4.4.2   Quadcopter Flight Scenarios

The proposed strategy is implemented on the quadrotor UAV with the same configuration as described previously. Note that the control gains are not changed in the case of disturbances imposed. The adjustment of control law lies in the outputs of DO and ESO. This is also the advantage of the MOBADC. Four groups of real-world flight tests (Fig. 4.4) are conducted in a laboratory environment as shown in Fig. 4.1. These experiments also demonstrate the advantages of the proposed method against both cable-suspended-payload disturbance and wind disturbance. The experimental parameters can be found in Appendix. The flight test videos can be found at https://youtu.be/6Fq1-aA-ZsM.

4.4.2.1   Test 1

To demonstrate the capability of the presented MOBADC against the cable-suspended-payload disturbance, a hovering flight under this disturbance is conducted first and the test can be checked in Fig. 4.4. In the course of the quadrotor UAV flying, the payload is static first and then forced to oscillate artificially. During the payload swings, the proposed DO based controller is triggered and the payload's oscillation amplitude remains the same during the whole process. Without loss of generality, the payload is forced to swing along y direction. σ_i is approximately equal to $\frac{1}{2\pi}\sqrt{g/L}$ s^{-1} where L is the length of the cable. Three main phases are included in this test, namely, *Phase I*: static payload using classical PID method only, *Phase II*: oscillating payload using classical PID method only, and *Phase III*: oscillating payload using the proposed controller with DO. The evolutions of the estimated disturbance of DO and the quadrotor's motion in y direction are shown in Figs. 4.6(a) and 4.6(b), respectively.

As seen from Fig. 4.6(a), compared with the static phase, the proposed DO is able to perceive and estimate the payload oscillating disturbance. This disturbance estimate is fed into our proposed position controller and then

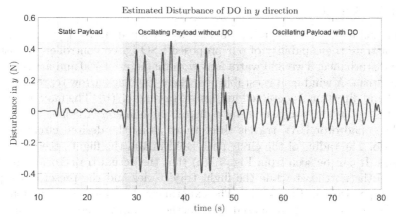

(a) Estimated disturbance of DO in y direction.

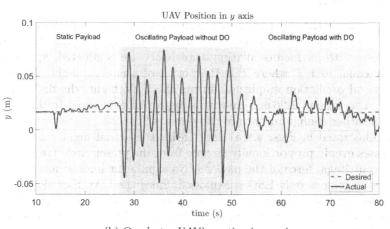

(b) Quadrotor UAV's motion in y axis.

FIGURE 4.6: **Hovering flight under payload oscillating disturbance of Test** 1: (a) and (b) depict the estimated disturbance and flight trajectory without/with DO in y direction, respectively. The dash and solid curves represent the desired and actual path.

the UAV's oscillation amplitude is reduced to around 1 cm as shown in Fig. 4.6(b). Obviously, the flight performance demonstrates that the proposed DO based controller is able to estimate the cable-suspended-payload disturbance efficiently and mitigate the UAV's hovering deviation.

4.4.2.2 Test 2

In this flight test, a circling flight under wind distance is carried out to demonstrate the capability of our proposed ESO based controller against the wind disturbance. Two 380 watts fans with diameter of 750 mm are set in the flying arena. A wind field is established where the black arrow represents wind speed and direction as depicted in Figs. 4.7(a) and 4.7(b). The maximum wind speed is up to 5 m/s measured by a digital anemometer AS8556.

The quadrotor UAV travels three times along the desired circle for each scenario. The radius of the circle path is 0.8 m and the flight speed is around 0.5 m/s. It can be seen from Fig. 4.7(a) that the actual trajectories are deviated to the northeast while the flight trajectories and the preset circle path overlap precisely in Fig. 4.7(b). The experimental performances demonstrate the effectiveness of the proposed ESO method against wind disturbance.

4.4.2.3 Test 3

To check whether the ESO only based controller can compensate the cable-suspended-payload disturbance for precise trajectory tracking, two circling flight tests with payload oscillating disturbance are conducted. σ_i is approximately equal to $1/T$ where T is the period of a circling flight. To increase the payload oscillation amplitude during the UAV flight, the flight speed is increased to around 1.26 m/s and σ_i is approximately equal to $0.25\,\mathrm{s}^{-1}$ and all of the other configurations are the same as Test 2.

As illustrated in Figs. 4.8(a) and 4.8(b), the actual flight trajectories in both cases evenly proportionally deviate from the preset circle trajectory due to the centrifugal force of the payload. No significant improvement has been achieved as long as only ESO is exploited. Since the UAV flies along a circle, the cable-suspended-payload disturbance can be modelled as a kind of disturbances with specific frequency. In such a situation, DO based controller should be involved.

4.4.2.4 Test 4

The results in three tests demonstrate the flight performances by using a single observer in different scenarios with only one disturbance. To demonstrate the capability of our proposed strategy against both payload oscillating disturbance and wind disturbance, four circling flight experiments under these two hybrid disturbances are carried out. The flight speeds in these tests are all set around 1.26 m/s, σ_i is approximately equal to $0.25\,\mathrm{s}^{-1}$ and the radius of the desired circle trajectory is 0.8 m as well. The quadrotor UAV flies three times along the desired path.

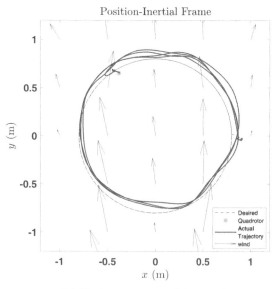

(a) Flight trajectory **without** ESO.

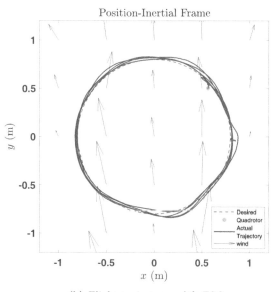

(b) Flight trajectory **with** ESO.

FIGURE 4.7: **Circling flight under wind disturbance of Test** 2: (a) and (b) depict the flight trajectories without/with ESO in x and y directions, respectively.

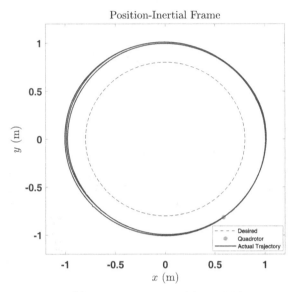

(a) Flight trajectory **without** ESO.

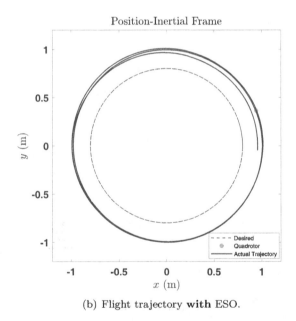

(b) Flight trajectory **with** ESO.

FIGURE 4.8: **Circling flight under payload disturbance of Test 3**: (a) and (b) depict the flight trajectories without/with ESO in x and y directions, respectively.

It can be observed from Fig. 4.9(a) and 4.9(b) that there exist significant trajectory deviations from the desired circle in these two scenarios. In comparison of Fig. 4.9(a), the shape of actual trajectories in Fig. 4.9(b) remains the same and bias is almost constant. This situation indicates that the ESO based controller has better wind resistance. Trajectories in Fig. 4.9(c) deviate less from the desired path when comparing to that in Fig. 4.9(b). It is depicted that the proposed DO based controller plays a significant role in attenuating the payload oscillating disturbance while the northward offset also shows that its wind resistance is not significant. Different from the above three cases, the

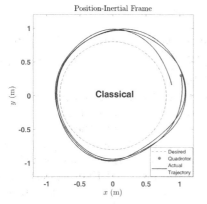

(a) Flight trajectories **without** any disturbance observers.

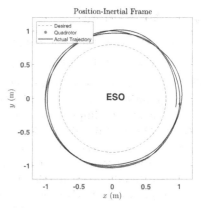

(b) Flight trajectories with **ESO only**.

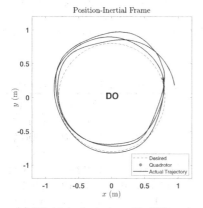

(c) Flight trajectories with **DO only**.

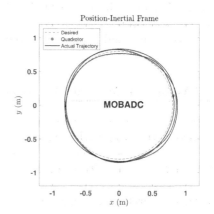

(d) Flight trajectories using **MOBADC**.

FIGURE 4.9: Circling flight under both payload oscillating and wind disturbances of Test 4: (a), (b), (c), and (d) depict the flight trajectories using classical method, ESO-based method, DO-based method and proposed MOBADC method, respectively.

flight trajectories in Fig. 4.9(d) have smaller trajectory deformation and offset, validating the effectiveness of our proposed strategy. The corresponding attitude information and motor control inputs by using MOBADC can be checked in Fig. 4.10. It can be seen from Fig. 4.10(a) that the attitude in response to three circling flights is relatively smooth. There exist some vibrations in roll angle when resisting wind disturbance and maintaining the desired trajectory. Under both payload oscillating disturbance and wind disturbance, the motors can be managed properly to handle these hybrid disturbances within an allowable range as depicted in Fig. 4.10(b). The flight tests results exemplify the applicability of the developed scheme in this study.

4.4.3 Assessment

To sum up, the statistical results (mean absolute error $\bar{\gamma} = \frac{1}{n}\sum_{i=1}^{n}\|\gamma_i - \gamma_{d,i}\|$ and standard deviation $s = \sqrt{\frac{1}{n-1}\sum_{i=1}^{n}(\|\gamma_i - \gamma_{d,i}\| - \bar{\gamma})^2}$) of the preceding experiments are summarized in Table 4.1.

The quantitative indices are listed in Table 4.1. It is confirmed that the DO based controller improves the mean error and STD of the trajectory tracking by 61.32% and 64.96% in Test 1, which coincides with the merits of the proposed DO in attenuating cable-suspended-payload disturbance. The ESO based method narrows the mean error and STD by 64.47% and 55.38% in Test 2, which validates the effectiveness of the proposed ESO against wind disturbance. However, there is no significant improvement in Test 3, which shows that the ESO only based method is not able to sufficiently mitigate the payload oscillating disturbance. In Test 4, the mean error and STD by using MOBADC can be improved by 76.70% and 71.14%, respectively. These results outperform the classical common-used PD controller, ESO and DO only based controller, and demonstrate the effectiveness of the proposed strategy.

TABLE 4.1: Mean absolute error and standard deviation. (Unit: m)

		Mean	STD
# **Test 1**-Hovering with payload	Classical	0.0181	0.0137
	DO	**0.0070**	**0.0048**
# **Test 2**-Circling with wind	Classical	0.0636	0.0316
	ESO	**0.0226**	**0.0141**
# **Test 3**-Circling with payload	Classical	**0.1442**	0.0361
	ESO	0.1930	**0.0141**
# **Test 4**-Circling with wind and payload	Classical	0.1502	0.0700
	ESO	0.2054	0.0205
	DO	0.0725	0.048
	MOBADC	**0.0350**	**0.0202**

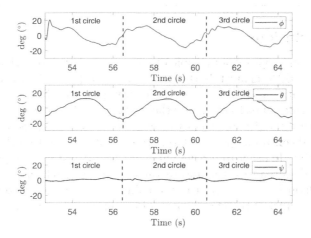

(a) Attitude angles by using MOBADC.

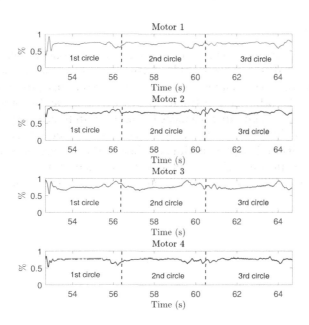

(b) Control inputs of motors during flight.

FIGURE 4.10: Attitude information and motor control inputs of Test 4 by using MOBADC.

4.5   Conclusions

This chapter proposed a multiple observers based anti-disturbance control for quadrotor UAV against both cable-suspended-payload disturbance and wind disturbance. A DO based controller was proposed in the translational control loop which aims to attenuate the payload oscillating disturbance. With respect to the wind disturbance, two ESOs were designed in the translational and rotational control loop respectively to mitigate its effect on the flight trajectory tracking. In order to further enhance the anti-disturbance performance for multiple disturbances, the anti-disturbance control scheme was proposed in position loop by combining the preceding two observers. Extensive flight experiments validated the effectiveness of our proposed system.

4.6   Notes

From the literature, two types of anti-disturbance control schemes can be potentially applied to quadrotor UAVs. One is to design a robust controller for the quadrotor UAV, making the closed-loop system insensitive to possible disturbances. The other is to estimate the disturbance, and subsequently attenuate or even eliminate disturbances by appropriately utilizing disturbance information. In this chapter, we adopted the second method. And both the simulation results and flight experiments showed the effectiveness of proposed strategy. In the future, safety flight control of quadrotor UAV in terms of centre-of-gravity shift together with motors' faults case can be investigated.

Chapter 5

Safety Control System Design of HGV Based on Adaptive TSMC

5.1 Introduction

Hypersonic gliding vehicle (HGV) is designed with the aerodynamic configuration of high lift-to-drag ratio, which can be launched into the sub-orbital trajectory either by a booster rocket or a reusable launch vehicle. Without any power, an HGV that can operate in near space with a speed of more than *Mach* 5, possesses a capacity for "extreme maneuvers." Benefited from the rapid response and flexible maneuverability, HGVs are recognized as a viable option of long-range delivery, remote rapid strike, and power projection.

One of the major problems is that large uncertainties and perturbations are inherent to the HGV model [108]. The HGV unique characteristics pose a severe challenge for the design of flight control systems. The vast majority of contributions focus on control methodologies, including adaptive control [109, 110], back-stepping and dynamic inversion control [111], linear parameter varying control [112], sliding mode control (SMC) [113], model predictive control [114], robust control [115], and adaptive continuous higher order sliding mode control (HOSMC) [116].

Requirements that drive the desire for HGVs include reliability and maintainability. By contrast, increasing complexity and automation make an HGV more vulnerable to component/system malfunctions across the flight envelope. The safety demand for hypersonic flights has spurred an interest in the control design. Takagi-Sugeno (T-S) fuzzy system and adaptive control have been exploited in the fault-tolerant control (FTC) design of hypersonic flight vehicles [117, 118]. Within the scheme [119], a group of local FTCs are synthesized in response to various faults, while Youla parameterizations are constructed to tolerate arbitrary switching actions among different fault modes of a hypersonic vehicle. A model predictive based FTC is investigated in Ref. [51], by which the reshaped reference can be tracked in the actuator faulty condition.

Two-loop controllers based on SMC are deployed to constitute an active FTC scheme for hypersonic vehicle attitude control. The resulting FTC system can guarantee the asymptotic output tracking in spite of actuator faults [120]. An FTC system on the basis of back-stepping and sliding mode technologies is applied for a hypersonic aircraft. The studied controller can guarantee the safety of the handicapped aircraft and the globally asymptotic tracking performance [121]. Recent effort [122] attempts to design an FTC system with combination of conventional SMC and nonlinear disturbance observer. Note that the actuator amplitude constraints and faults are taken into account over the FTC design phase. Furthermore, terminal SMC (TSMC) techniques are adopted to advance the state of the art of FTC. TSMC can not only possess strong robustness on uncertain dynamics similar to linear SMC, but also guarantee the finite-time convergence of tracking error [123, 124, 125]. Two dynamic TSMCs are designed with respect to the inner and outer loops, handling actuator faults of a hypersonic vehicle [126]. An FTC strategy is determined by resorting to TSMC approach, ensuring that velocity and altitude track the reference signals in finite time after occurrence of actuator malfunctions [127]. The work in Ref. [128], which develops a passive FTC based on TSMC technique, focuses on enhancing the convergence rate.

Despite that previous studies have gained various degrees of success in addressing HGV safety control issues, there still exist some challenges.

1) As mentioned in Chapter 1, the amount of fault recovery time, which solely relies on the operating condition and the fault characteristics, is very limited for safety-critical plants [4, 129]. More particularly, the safety restrictions imposed on HGV inputs and outputs may be violated, if faults cannot be accommodated within the allowable amount of time. From this perspective, more emphasis on the HGV safety control design with a timely manner needs to be placed.

2) In terms of the time-scale separation principle, it is common practice to approach the HGV safety control problem by independent design of fast inner-loop and slow outer-loop dynamics [130, 131]. In the outer-loop, the angular rate profiles, which are regarded as virtual control signals to the inner-loop, are produced by the kinematics equation of angular motion and the SMC. With respect to the inner loop, another SMC is synthesized such that the commanded angular rate profiles are tracked. Roll, pitch, and yaw torque commands generated by the inner-loop are then allocated into control surface deflection commands. However, how to guarantee the finite-time stability of the overall system is an open issue.

3) TSMC is exploited for stabilizing the HGV subject to faults and uncertainties. Nonetheless, in the most of resulting TSMC approaches, a multi-input control problem with m control inputs is transformed into a decoupled problem involving m single input control structures. This type of approach may not be effective due to strong couplings in HGV aerodynamics.

In an attempt to tackle the above-mentioned issues, a TSMC based safety control design approach is proposed against HGV actuator faults and model uncertainties, with particular attention devoted to achieving multivariable design in a composite-loop. The major contributions are briefly outlined as follows.

1) Due to lack of wind tunnel facility and flight test experiments, a partial knowledge of the aerodynamic derivatives of hypersonic vehicles is present. The control input matrix in any HGV is composed of control moment coefficients, which are extremely difficult to accurately obtain in comparison of conventional aircraft. Hence, multiplicative uncertainty exists in the HGV control input matrix, inducing a great challenge of control design especially in the event of actuator faults. In this study, the cases of HGV actuator malfunctions and multiplicative uncertainty in control input matrix are simultaneously considered at the safety control design stage. To the best of the authors' knowledge, there are few papers focusing on this aspect.

2) In most of the existing literature, control design of hypersonic vehicles is divided into the inner loop and outer loop design (named dual-loop design) based on time-scale separation principle. However, this type of design cannot ensure the stability of the overall closed-loop system. This study establishes a control-oriented model by integrating the HGV attitude kinematic and dynamic equations. Subsequently, a composite-loop design for HGV attitude tracking control under actuator faults is developed. The finite-time stability of the closed-loop system can be guaranteed from a theoretical perspective.

3) A finite-time multivariable TSMC approach based on homogeneity is exploited in the safety control design. With consideration of HGV actuator malfunctions and model uncertainties, a novel integral terminal sliding mode surface is established by introducing the fractional power integral terms. The resulting safety control can ensure the finite-time stability of the HGV, when actuator faults and model uncertainties exist. Moreover, the multivariable integral TSMC formed by vector expression, which is driven directly from the sliding mode reachability condition, is incorporated in the HGV safety control design. This feature is of significance in the sense that the strong couplings are inherent to HGV aerodynamics.

The rest of this chapter is organized as follows. The concepts of finite-time stable system and homogeneity are described in Chapter 5.2. The HGV aerodynamics, actuator fault model, and problem statement are given in Chapter 5.3. The control-oriented model is presented in Chapter 5.4. The HGV safety control scheme is proposed against actuator faults and model uncertainties in Chapter 5.5. In Chapter 5.6, the performance of the developed safety control is evaluated through simulations of a full nonlinear model of the HGV dynamics. Chapter 5.7 includes a discussion of the conclusions.

5.2   Preliminaries

A brief description of finite-time stability and homogeneity is presented, serving as a foundation of the HGV safety control design.

Consider the system:

$$\dot{\xi} = f(\xi), \ f(0) = 0, \ \xi \in R^n, \ \xi(0) = x_0, \tag{5.1}$$

where $f : D \to R^n$ is continuous on an open neighborhood D of the origin $\xi = 0$. The equilibrium $\xi = 0$ of Eq. (5.1) is finite-time convergent if there are an open neighborhood $U \subseteq D$ of the origin and a function $T_\xi : U \backslash \{0\} \to (0, \infty)$, such that $\forall \xi_0 \in U$, the solution trajectories $\xi(t, \xi_0)$ of Eq. (5.1) starting from the initial point $\xi_0 \in U \backslash \{0\}$ is well-defined and unique in forward time for $t \in [0, T_\xi(\xi_0))$, and $\lim_{t \to T_\xi(\xi_0)} \xi(t, \xi_0) = 0$. Here $T_\xi(\xi_0)$ is called the settling time (of the initial state ξ_0). The equilibrium of Eq. (5.1) is finite-time stable if it is Lyapunov stable and finite-time convergent. When $U = D = R^n$, then the origin is in globally finite-time stable equilibrium.

Definition 5.1. *Let dilation* $(r_1, \cdots, r_n) \in R^n$ *with* $r_i > 0$, $i = 1, \cdots, n$. *Let* $f(\xi) = [f_1(\xi), \cdots, f_n(\xi)]^T$ *be a continuous vector field.* $f(\xi)$ *is recognized to be homogeneous of degree* $d \in R$ *with respect to dilation* $(r_1, \cdots, r_n)$ *if, for any given* $\varepsilon > 0$,

$$f_i(\varepsilon^{r_1} \xi_1, \cdots, \varepsilon^{r_n} \xi_n) = \varepsilon^{d + r_i} f_i(\xi), \quad i = 1, \ldots n, \forall \xi \in R^n. \tag{5.2}$$

System Eq. (5.1) is said to be homogeneous if $f(\xi)$ *is homogeneous.*

Lemma 5.1. *[132] The continuous system Eq. (5.1) is named globally finite-time stable if it is globally asymptotical stable and locally homogeneous of degree* $d < 0$.

5.3   Mathematical Model of a HGV

5.3.1   Nonlinear HGV Model

The HGV model is based on the assumption of a rigid vehicle structure, a flat, non-rotating Earth and uniform gravitational field. In the following, the kinematic model and dynamic model of a HGV are described, respectively. The inertial position coordinates are described as:

$$\begin{cases} \dot{x} = V \cos \gamma \cos \chi \\ \dot{y} = V \cos \gamma \sin \chi \\ \dot{z} = -V \sin \gamma \end{cases}, \tag{5.3}$$

where x, y, and z represent the positions with respect to x-, y-, and z-directions of the Earth-fixed reference frame, respectively. V stands for the total velocity, γ and χ denote the flight-path angle and the heading angle, respectively.

The force equations are described as:

$$
\begin{cases}
\dot{V} = -g\sin\gamma - \dfrac{QS_rC_D}{m} \\[2mm]
\dot{\chi} = \dfrac{QS_r}{mV\cos\gamma}(C_L\sin\mu + C_Y\cos\mu) \\[2mm]
\dot{\gamma} = -\dfrac{g}{V}\cos\gamma + \dfrac{QS_r}{mV}(C_L\cos\mu - C_Y\sin\mu)
\end{cases}
\qquad (5.4)
$$

where μ is the bank angle, g is the gravitational constant, Q is the dynamic pressure, S_r is the reference area, m is the HGV mass, C_L, C_D, and C_Y are the aerodynamic coefficients with respect to lift, drag, and side force, respectively.

The model of attitude is written as:

$$
\begin{cases}
\dot{\mu} = \sec\beta(p\cos\alpha + r\sin\alpha) \\[2mm]
\quad + \dfrac{QS_rC_L}{mV}(\tan\gamma\sin\mu + \tan\beta) \\[2mm]
\quad + \dfrac{QS_rC_Y}{mV}\tan\gamma\cos\mu - \dfrac{g}{V}\cos\gamma\cos\mu\tan\beta \\[2mm]
\dot{\alpha} = q - \tan\beta(p\cos\alpha + r\sin\alpha) \\[2mm]
\quad + \dfrac{1}{mV\cos\beta}(mg\cos\gamma\cos\mu - QS_rC_L) \\[2mm]
\dot{\beta} = -r\cos\alpha + p\sin\alpha \\[2mm]
\quad + \dfrac{1}{mV}(QS_rC_Y + mg\cos\gamma\sin\mu)
\end{cases}
\qquad (5.5)
$$

where α and β denote the angle of attack (AOA) and sideslip angle, respectively.

The model of angular velocities is given as:

$$
\begin{cases}
\dot{p} = \dfrac{l_A + (I_{yy} - I_{zz})qr}{I_{xx}} \\[2mm]
\dot{q} = \dfrac{m_A + (I_{zz} - I_{xx})pr}{I_{yy}} \\[2mm]
\dot{r} = \dfrac{n_A + (I_{xx} - I_{yy})pq}{I_{zz}}
\end{cases}
\qquad (5.6)
$$

where p, q, and r are roll, pitch, and yaw angular rates, respectively. l_A, m_A, and n_A denote the roll, pitch, and yaw moments, while I_{xx}, I_{yy}, and I_{zz} represent the moments of inertia.

The aerodynamics forces L, D, and Y are represented as:

$$
\begin{cases}
L = QS_rC_L \\[2mm]
D = QS_rC_D \\[2mm]
Y = QS_rC_Y
\end{cases}
\qquad (5.7)
$$

where $C_L = C_{L,clean} + C_{L,\delta_a}\delta_a + C_{L,\delta_e}\delta_e$, $C_D = C_{D,clean} + C_{D,\delta_a}\delta_a + C_{D,\delta_e}\delta_e + C_{D,\delta_r}\delta_r$, and $C_Y = C_{Y,\beta} + C_{Y,\delta_a}\delta_a + C_{Y,\delta_e}\delta_e + C_{Y,\delta_r}\delta_r$. δ_a, δ_e, and δ_r are the so-called control deflections of the aileron, elevator, and rudder, respectively.

The rolling, pitching, and yawing moments are:

$$\begin{cases} l_A = QbS_rC_l \\ m_A = QcS_rC_m - x_{cg}(-D\sin\alpha - L\cos\alpha), \\ n_A = QbS_rC_n + x_{cg}Y \end{cases} \tag{5.8}$$

where b is the span of the HGV, c is the mean aerodynamic chord, x_{cg} is the distance between the centroid and reference moment along x body-axis. The corresponding coefficients are calculated as: $C_l = C_{l,\beta}\beta + C_{l,\delta_a}\delta_a + C_{l,\delta_e}\delta_e + C_{l,\delta_r}\delta_r + C_{l,r}\frac{rb}{2V} + C_{l,p}\frac{pb}{2V}$, $C_m = C_{m,clean} + C_{m,\delta_a}\delta_a + C_{m,\delta_e}\delta_e + C_{m,\delta_r}\delta_r + C_{m,q}\frac{qc}{2V}$, and $C_n = C_{n,\beta}\beta + C_{n,\delta_a}\delta_a + C_{n,\delta_e}\delta_e + C_{n,\delta_r}\delta_r + C_{n,p}\frac{pb}{2V} + C_{n,r}\frac{rb}{2V}$.

5.3.2 Actuator Fault Model

When actuation systems work under a normal condition, appropriate aerodynamic forces and moments are produced. The required HGV maneuver can be thereby accomplished with a baseline/nominal controller. If the HGV encounters actuator malfunctions, the nominal controller's attempts to maintain the expected maneuver may be futile and the flight safety can be jeopardized [30, 133]. Gain fault and bias fault are the faults commonly appearing on flight actuators. The actuator fault model is generally formed as:

$$\boldsymbol{u}_F = \boldsymbol{\Lambda}\boldsymbol{u} + \boldsymbol{\rho}, \tag{5.9}$$

where $\boldsymbol{\Lambda} = \text{diag}\{\lambda_1, \lambda_2, \lambda_3\}$ represents the gain fault, $\boldsymbol{\rho} = [\rho_1, \rho_2, \rho_3]^T$ denotes the bias fault, and $\boldsymbol{u} = [\delta_a, \delta_e, \delta_r]^T$.

Remark 5.1. *It is reported in Ref. [30] that the leakage of hydraulic fluid can be the root cause of the degradation of the actuator effectiveness. Therefore, $\boldsymbol{\Lambda} = \text{diag}\{\lambda_1, \lambda_2, \lambda_3\}$ in Eq. (5.9) is used to describe the effectiveness of the HGV actuators, where $0 < \lambda_1, \lambda_2, \lambda_3 \leq 1$. In addition, the sensor fault in an actuator system can result in the actuator bias faults. To be more specific, if the amplitude sensor encounters a bias fault, the measured amplitude is the actual amplitude plus the bias value. As a consequence, the sensed amplitude is forced to be equal to the referenced signal. However, the actual value of the actuator amplitude is deviated from the expected value. Hence, $\boldsymbol{\rho} = [\rho_1, \rho_2, \rho_3]^T$ is adopted in Eq. (5.9) to describe the bias faults of the aileron, elevator, and rudder, respectively.*

5.3.3 Problem Statement

The purpose is to develop a safety control scheme based on adaptive multivariable integral TSMC such that:

1) The deleterious effects of HGV actuator faults can be compensated within a finite amount of time, thus:

$$\lim_{t \to t_f} |\mu - \mu_d| = 0, \ \lim_{t \to t_f} |\alpha - \alpha_d| = 0, \ \lim_{t \to t_f} |\beta - \beta_d| = 0, \tag{5.10}$$

where t_f is the finite time, μ_d, α_d, and β_d correspond to the reference commands of the bank angle, AOA, and sideslip angle, respectively;

2) A composite-loop design for HGV attitude tracking control under actuator faults can be achieved, without the need of dividing the HGV dynamics into the inner-loop and outer-loop; and

3) Multivariable design can be integrated into the safety control.

5.4 Control-Oriented Model

The establishment of the control-oriented model is presented, which provides a basis of the composite-loop design of the HGV safety control.

For the safety flight control of the HGV, define $x_1 = [\mu, \alpha, \beta]^T$ and $x_2 = [p, q, r]^T$. In accordance with Eq. (5.5), one can obtain:

$$\begin{cases} \dot{\mu} = \sec \beta (p \cos \alpha + r \sin \alpha) + f_\mu \\ \dot{\alpha} = q - \tan \beta (p \cos \alpha + r \sin \alpha) + f_\alpha \ , \\ \dot{\beta} = -r \cos \alpha + p \sin \alpha + f_\beta \end{cases} \tag{5.11}$$

where

$$\begin{cases} f_\mu = \dfrac{QS_r C_L}{mV}(\tan \gamma \sin \mu + \tan \beta) \\ \qquad + \dfrac{QS_r C_Y}{mV} \tan \gamma \cos \mu - \dfrac{g}{V} \cos \gamma \cos \mu \tan \beta \\ f_\alpha = \dfrac{1}{mV \cos \beta}(mg \cos \gamma \cos \mu - QS_r C_L) \\ f_\beta = \dfrac{1}{mV}(QS_r C_Y \cos \beta + mg \cos \gamma \sin \mu) \end{cases} . \tag{5.12}$$

By recalling the definitions $x_1 = [\mu, \alpha, \beta]^T$ and $x_2 = [p, q, r]^T$, Eqs. (5.11)–(5.12) can be therefore expressed as:

$$\dot{x}_1 = f_1 + g_1 x_2, \tag{5.13}$$

where $f_1 = [f_\mu, f_\alpha, f_\beta]^T$, and

$$g_1 = \begin{bmatrix} \sec \beta \cos \alpha & 0 & \sec \beta \sin \alpha \\ -\tan \beta \cos \alpha & 1 & -\tan \beta \sin \alpha \\ \sin \alpha & 0 & -\cos \alpha \end{bmatrix} . \tag{5.14}$$

Moreover, combining the angular rate dynamics of Eq. (5.6) and aerodynamic moments of Eq. (5.8) gives:

$$\dot{p} = \left(QbS_r \left(C_{l,\beta}\beta + C_{l,r} \left(\frac{rb}{2V} \right) + C_{l,p} \left(\frac{pb}{2V} \right) \right) \right) / I_{xx}$$
$$+ \left((I_{yy} - I_{zz}) qr \right) / I_{xx}$$
$$+ \left(QbS_r \left(C_{l,\delta a}\delta_a + C_{l,\delta e}\delta_e + C_{l,\delta r}\delta_r \right) \right) / I_{xx}. \tag{5.15}$$

By defining $f_p = QbS_r \left(C_{l,\beta}\beta + C_{l,r} \left(\frac{rb}{2V} \right) + C_{l,p} \left(\frac{pb}{2V} \right) \right) / I_{xx} + ((I_{yy} - I_{zz})qr)$
$/I_{xx}$, Eq. (5.15) can be recast in the form:

$$\dot{p} = f_p + \left[\frac{QbS_r C_{l,\delta a}}{I_{xx}}, \frac{QbS_r C_{l,\delta e}}{I_{xx}}, \frac{QbS_r C_{l,\delta r}}{I_{xx}} \right] \cdot \begin{bmatrix} \delta_a & \delta_e & \delta_r \end{bmatrix}^T. \tag{5.16}$$

Similarly, the pitch angular rate dynamics is represented as:

$$\dot{q} = \left(QcS_r \left(C_{m,\text{ clean}} + C_{m,q} \left(\frac{qc}{2V} \right) \right) \right) / I_{yy}$$
$$+ Qx_{cg}S_r \left(C_{D,\text{ clean}} \sin\alpha + C_{L,\text{ clean}} \cos\alpha \right) / I_{yy}$$
$$+ \left(QcS_r \left(C_{m,\delta a}\delta_a + C_{m,\delta e}\delta_e + C_{m,\delta r}\delta_r \right) \right) / I_{yy}$$
$$+ \left(Qx_{cg}S_r \left((C_{D,\delta a}\sin\alpha + C_{L,\delta a}\cos\alpha)\,\delta_a \right) \right) / I_{yy}$$
$$+ \left(Qx_{cg}S_r \left((C_{D,\delta e}\sin\alpha + C_{L,\delta a}\cos\alpha)\,\delta_e \right) \right) / I_{yy}$$
$$+ \left(Qx_{cg}S_r \left(C_{D,\delta r}\sin\alpha\delta_r \right) \right) / I_{yy}$$
$$+ \left((I_{zz} - I_{xx})\,pr \right) / I_{yy}. \tag{5.17}$$

As long as letting $f_q = (QcS_r(C_{m,clean} + C_{mq}(\frac{qc}{2V})))/I_{yy} + (Qx_{cg}S_r(C_{D,clean}$
$\sin\alpha + C_{L,clean}\cos\alpha))/I_{yy} + ((I_{zz} - I_{xx})pr)/I_{yy}$, Eq. (5.17) can be simplified
as:

$$\dot{q} = f_q + \begin{bmatrix} \dfrac{QcS_r C_{m,\delta a} + Qx_{cg}S_r(C_{D,\delta a}\sin\alpha + C_{L,\delta a}\cos\alpha)}{I_{yy}} \\[2mm] \dfrac{QcS_r C_{m,\delta e} + Qx_{cg}S_r(C_{D,\delta e}\sin\alpha + C_{L,\delta e}\cos\alpha)}{I_{yy}} \\[2mm] \dfrac{QcS_r C_{m,\delta r} + Qx_{cg}S_r C_{D,\delta r}\sin\alpha}{I_{yy}} \end{bmatrix}^T \times \begin{bmatrix} \delta_a \\ \delta_e \\ \delta_r \end{bmatrix}. \tag{5.18}$$

The yaw angular rate dynamics is described as:

$$\dot{r} = f_r + \begin{bmatrix} \dfrac{QbS_r C_{n,\delta a} + Qx_{cg}S_r C_{Y,\delta a}}{I_{zz}} \\[2mm] \dfrac{QbS_r C_{n,\delta e} + Qx_{cg}S_r C_{Y,\delta e}}{I_{zz}} \\[2mm] \dfrac{QbS_r C_{n,\delta r} + Qx_{cg}S_r C_{Y,\delta r}}{I_{zz}} \end{bmatrix}^T \begin{bmatrix} \delta_a \\ \delta_e \\ \delta_r \end{bmatrix}, \tag{5.19}$$

where $f_r = QbS_r(C_{n,\beta}\beta + C_{np}(\frac{pb}{2V}) + C_{nr}(\frac{rb}{2V}))\Big/I_{zz} + Qx_{cg}S_rC_{Y,\beta}\beta/I_{zz},$
$+(I_{xx} - I_{yy})pq/I_{zz}.$

According to Eqs. (5.16), (5.18), and (5.19), one can obtain:

$$\dot{x}_2 = f_2 + g_2 u, \qquad (5.20)$$

$$g_2 = \begin{bmatrix} \dfrac{QbS_rC_{l,\delta a}}{I_{xx}} & \dfrac{QbS_rC_{l,\delta e}}{I_{xx}} \\[2mm] \dfrac{QcS_rC_{m,\delta a} + Qx_{cg}S_r(}{C_{D,\delta a}\sin\alpha + C_{L,\delta a}\cos\alpha)}{I_{yy}} & \dfrac{QcS_rC_{m,\delta e} + Qx_{cg}S_r(}{C_{D,\delta e}\sin\alpha + C_{L,\delta e}\cos\alpha)}{I_{yy}} \\[2mm] \dfrac{QbS_rC_{n,\delta a} + Qx_{cg}S_rC_{Y,\delta a}}{I_{zz}} & \dfrac{QbS_rC_{n,\delta e} + Qx_{cg}S_rC_{Y,\delta e}}{I_{zz}} \end{bmatrix}$$

$$\begin{matrix} \dfrac{QbS_rC_{l,\delta r}}{I_{xx}} \\[2mm] \dfrac{QcS_rC_{m,\delta r} + Qx_{cg}S_rC_{D,\delta r}\sin\alpha}{I_{yy}} \\[2mm] \dfrac{QbS_rC_{n,\delta r} + Qx_{cg}S_rC_{Y,\delta r}}{I_{zz}} \end{matrix} \Bigg],$$

$$(5.21)$$

where $f_2 = [f_p, f_q, f_r]^T$, $u = [\delta_a, \delta_e, \delta_r]^T$.

Combining Eqs. (5.13) and (5.20) gives:

$$\begin{cases} \dot{x}_1 = f_1 + g_1 x_2 \\ \dot{x}_2 = f_2 + g_2 u \end{cases}. \qquad (5.22)$$

Differentiating Eq. (5.13) and recalling Eq. (5.20) achieve:

$$\ddot{x}_1 = \dot{f}_1 + \dot{g}_1 x_2 + g_1 f_2 + g_1 g_2 u. \qquad (5.23)$$

The control-oriented model with actuator anomalies and model uncertainties is built as follows. Firstly, Eq. (5.22) can be expressed as:

$$\ddot{x}_1 = \dot{f}_1 + F(x_1, x_2) + G(x_1, x_2)u, \qquad (5.24)$$

where $F(x_1, x_2) = \dot{g}_1 x_2 + g_1 f_2$ and $G(x_1, x_2) = g_1 g_2$. In addition, $F(x_1, x_2)$ contains two terms:

$$F = F_n + \Delta_F, \qquad (5.25)$$

where F_n and Δ_F denote the nominal portion and the uncertain portion of F, respectively. $G(x_1, x_2)$ can be specified in a manner similar to $F(x_1, x_2)$:

$$G = G_n + \Delta_G. \qquad (5.26)$$

The nominal term of G is G_n, which solely relies on the known portions of g_1 and g_2. With respect to the studied HGV, $\det(g_1) = -sec\beta$. One can obtain that g_1 is invertible if β does not equal to $\pm\pi/2$. Focusing on the known portion of g_2, it can be regarded as control allocation matrix which is invertible in HGV flight envelopes. Therefore, the nominal portion G_n is invertible.

Consequently, Eq. (5.23) can be further written as:

$$\ddot{x}_1 = \dot{f}_1 + F_n + \Delta_F + (G_n + \Delta_G)u. \qquad (5.27)$$

By accounting for the gain and the bias faults in actuators as Eq. (5.9), one can render:

$$\ddot{x}_1 = \dot{f}_1 + F_n + \Delta_F + (G_n + \Delta_G)(\Lambda u + \rho)$$

$$= \dot{f}_1 + F_n + \Delta_F + (G_n + \Delta_G)\rho \qquad (5.28)$$

$$+ (G_n + \Delta_G\Lambda + G_n(\Lambda - I))u,$$

where I is a 3×3 identity matrix.

Assumption 5.1. *It is assumed that the boundedness of $\dot{f}_\mu$, $\dot{f}_\alpha$, and $\dot{f}_\beta$ is associated with the norm of system states. It can be further assumed that:*

$$\begin{cases} \left\| \dot{f}_1 + \Delta_F + (G_n + \Delta_G)\rho \right\| \leq \varepsilon_1 + \varepsilon_2 \|x\| \\ \left\| (\Delta_G\Lambda + G_n(\Lambda - I))G_n^{-1} \right\| \leq \varepsilon_3 < 1 \end{cases}, \qquad (5.29)$$

where ε_1, ε_2, and ε_3 are positive scalars.

Remark 5.2. $f_1 = [f_\mu, f_\alpha, f_\beta]^T$ *can be seen as an impact term of trajectory on the HGV attitude. Since the attitude dynamics is much faster than the translation motion, the values of f_μ, f_α, and f_β are usually small. As can be seen from (5.29), the lumped additive uncertainty term $\dot{f}_1 + \Delta_F + (G_n + \Delta_G)\rho$ is dependent on the system states. $\dot{f}_\mu$ is not only greatly dependent on γ, μ, α, β, and V, but also on p, q, and r. Essentially, the HGV angles as well as the HGV velocity are bounded in typical HGV flight envelopes. Therefore, the bound of $\dot{f}_\mu$ is closely related to the norm of HGV states if $\beta \approx 0$, $\gamma \neq 90°$, and $V \neq 0$. The similar assumption can be applied to the boundedness of $\dot{f}_\alpha$ and $\dot{f}_\beta$.*

Remark 5.3. *The second inequality in (5.29) is essential such that the control signal $G_n u$ dominates the uncertain vector function $(\Delta_G\Lambda + G_n(\Lambda - I))u$, which is induced by the actuator anomalies and the control input matrix uncertainty. This condition in turn ensures that the actuators configured are capable of addressing the HGV uncertainty and fault issues. In addition, G_n and Δ_G solely rely on HGV angles including α and β which are bounded in typical flight envelopes, instead of the angular velocities p, q, and r. Hence, it is assumed that the second term of (5.29) is bounded.*

Remark 5.4. *Time-scale separation of the independent inner-loop and outer-loop designs, stemming from Eq. (5.22), is typically enforced in most of the design approaches. It is difficult to guarantee the finite-time stability of the overall closed-loop system. By contrast, Eq. (5.28) is integrated by Eqs. (5.13) and (5.20), which provides the basis of the proposed composite-loop design.*

5.5   Safety Control System Design of a HGV against Faults and Uncertainties

The system depicted in Fig. 5.1 is composed by the guidance and control units. Generally, the guidance system generates the commands of the bank angle, AOA, and sideslip angle. The safety control scheme that is the main focus of this study outputs the actuator commands necessary to track the desired attitude and to handle actuator malfunctions. Two problems are addressed in the following. The first is the composite-loop design of safety control problem: construct the HGV safety control law against actuator malfunctions and model uncertainties, using the multivariable integral TSMC technique. The second is the problem of selecting the control parameters within the developed safety control scheme: determine the control parameters by exploring the adaptive tuning method.

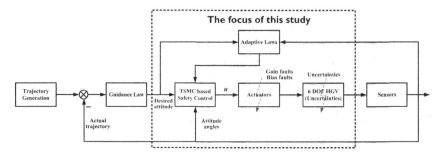

FIGURE 5.1: Schematic illustration of the studied HGV safety control.

5.5.1   Multivariable TSMC

Define the tracking error vector as:

$$\boldsymbol{\sigma} = \begin{bmatrix} \sigma_\mu \\ \sigma_\alpha \\ \sigma_\beta \end{bmatrix} = \begin{bmatrix} \mu - \mu_d \\ \alpha - \alpha_d \\ \beta - \beta_d \end{bmatrix}. \tag{5.30}$$

The multivariable integral terminal sliding mode manifold is defined as:

$$\boldsymbol{S} = \begin{bmatrix} S_\mu \\ S_\alpha \\ S_\beta \end{bmatrix} = \dot{\boldsymbol{\sigma}} + \int_0^t k_1 \|\boldsymbol{\sigma}\|^{r_1} \frac{\boldsymbol{\sigma}}{\|\boldsymbol{\sigma}\|} + k_2 \|\dot{\boldsymbol{\sigma}}\|^{r_2} \frac{\dot{\boldsymbol{\sigma}}}{\|\dot{\boldsymbol{\sigma}}\|} d\tau, \tag{5.31}$$

where $r_2 \in (0,1)$, $r_1 = 2r_2/(2 - r_2)$, and $k_1, k_2 > 0$. The multivariable integral TSMC based FTC aims at steering the tracking error vector $\boldsymbol{\sigma}$ to the origin along $\boldsymbol{S} = \boldsymbol{0}$ in finite time, under actuator faults and model uncertainties.

Theorem 5.1. *The HGV safety control law is formed as:*

$$u = u_b + u_d, \tag{5.32}$$

where

$$u_b = -G_n^{-1}\left[F_n - x_{1,d}^{(2)} + k_1\|\sigma\|^{r_1}\frac{\sigma}{\|\sigma\|} + k_2\|\dot\sigma\|^{r_2}\frac{\dot\sigma}{\|\dot\sigma\|}\right], \tag{5.33}$$

$$u_d = \begin{cases} -G_n^{-1}\left(c_1 + c_2\|x\| + c_3\left\|F_n - x_{1,d}^{(2)} + k_1\|\sigma\|^{r_1}\frac{\sigma}{\|\sigma\|}\right. \right. \\ \qquad \left. \left. + k_2\|\dot\sigma\|^{r_2}\frac{\dot\sigma}{\|\dot\sigma\|}\right\| + \eta\right)\frac{S}{\|S\|}, & S \neq 0 \\ 0, & S = 0 \end{cases} \tag{5.34}$$

In Eqs. (5.32)–(5.34), $x_{1,d}^{(2)}$ denotes the second time derivative of the desired x_1 (i.e., $x_{1,d} = [\mu_d, \alpha_d, \beta_d]^T$). c_1, c_2, c_3, and η are the design parameters which are chosen as:

$$c_1 = \frac{\varepsilon_1}{1 - \varepsilon_3}, c_2 = \frac{\varepsilon_2}{1 - \varepsilon_3}, c_3 = \frac{\varepsilon_3}{1 - \varepsilon_3}, \eta > 0. \tag{5.35}$$

Therefore, the designed safety control law ensures that the tracking error vector σ is driven to the origin along $S = 0$ in finite time regardless of actuator faults and model uncertainties.

Proof. Consider the following Lyapunov function:

$$V_1 = \frac{S^T S}{2}. \tag{5.36}$$

The time derivative of V_1 for $S \neq 0$ is:

$$\begin{aligned} \dot V_1 &= S^T\left(\ddot\sigma + k_1\|\sigma\|^{r_1}\frac{\sigma}{\|\sigma\|} + k_2\|\dot\sigma\|^{r_2}\frac{\dot\sigma}{\|\dot\sigma\|}\right) \\ &= S^T\left(F_n + \dot f_1 + \Delta_F + (G_n + \Delta_G)\rho \right. \\ &\qquad + (G_n + \Delta_G\Lambda + G_n(\Lambda - I))u - x_{1,d}^{(2)} \\ &\qquad \left. + k_1\|\sigma\|^{r_1}\frac{\sigma}{\|\sigma\|} + k_2\|\dot\sigma\|^{r_2}\frac{\dot\sigma}{\|\dot\sigma\|}\right). \end{aligned} \tag{5.37}$$

Substituting Eqs. (5.33)–(5.34) into Eq. (5.37) yields:

$$
\begin{aligned}
\dot{V}_1 &= \boldsymbol{S}^T(\boldsymbol{F}_n + \dot{\boldsymbol{f}}_1 + \boldsymbol{\Delta}_F + (\boldsymbol{G}_n + \boldsymbol{\Delta}_G)\boldsymbol{\rho} \\
&\quad - (\boldsymbol{G}_n + \boldsymbol{\Delta}_G\boldsymbol{\Lambda} + \boldsymbol{G}_n(\boldsymbol{\Lambda} - \boldsymbol{I}))\boldsymbol{G}_n^{-1}(\boldsymbol{F}_n - \boldsymbol{x}_{1,d}^{(2)}) \\
&\quad + k_1 \|\boldsymbol{\sigma}\|^{r_1} \frac{\boldsymbol{\sigma}}{\|\boldsymbol{\sigma}\|} + k_2 \|\dot{\boldsymbol{\sigma}}\|^{r_2} \frac{\dot{\boldsymbol{\sigma}}}{\|\dot{\boldsymbol{\sigma}}\|} + (c_1 + c_2 \|\boldsymbol{x}\| \\
&\quad + c_3 \left\| \boldsymbol{F}_n - \boldsymbol{x}_{1,d}^{(2)} + k_1 \|\boldsymbol{\sigma}\|^{r_1} \frac{\boldsymbol{\sigma}}{\|\boldsymbol{\sigma}\|} + k_2 \|\dot{\boldsymbol{\sigma}}\|^{r_2} \frac{\dot{\boldsymbol{\sigma}}}{\|\dot{\boldsymbol{\sigma}}\|} \right\| \\
&\quad + \eta)\frac{\boldsymbol{S}}{\|\boldsymbol{S}\|}) + k_1 \|\boldsymbol{\sigma}\|^{r_1} \frac{\boldsymbol{\sigma}}{\|\boldsymbol{\sigma}\|} + k_2 \|\dot{\boldsymbol{\sigma}}\|^{r_2} \frac{\dot{\boldsymbol{\sigma}}}{\|\dot{\boldsymbol{\sigma}}\|} - \boldsymbol{x}_{1,d}^{(2)}) \\
&= \boldsymbol{S}^T(\dot{\boldsymbol{f}}_1 + \boldsymbol{\Delta}_F + (\boldsymbol{G}_n + \boldsymbol{\Delta}_G)\boldsymbol{\rho} - (c_1 + c_2 \|\boldsymbol{x}\| \\
&\quad + c_3 \left\| \boldsymbol{F}_n - \boldsymbol{x}_{1,d}^{(2)} + k_1 \|\boldsymbol{\sigma}\|^{r_1} \frac{\boldsymbol{\sigma}}{\|\boldsymbol{\sigma}\|} + k_2 \|\dot{\boldsymbol{\sigma}}\|^{r_2} \frac{\dot{\boldsymbol{\sigma}}}{\|\dot{\boldsymbol{\sigma}}\|} \right\| \\
&\quad + \eta)\frac{\boldsymbol{S}}{\|\boldsymbol{S}\|} - (\boldsymbol{\Delta}_G\boldsymbol{\Lambda} + \boldsymbol{G}_n(\boldsymbol{\Lambda} - \boldsymbol{I}))\boldsymbol{G}_n^{-1}(\boldsymbol{F}_n - \boldsymbol{x}_{1,d}^{(2)} \\
&\quad + k_1 \|\boldsymbol{\sigma}\|^{r_1} \frac{\boldsymbol{\sigma}}{\|\boldsymbol{\sigma}\|} + k_2 \|\dot{\boldsymbol{\sigma}}\|^{r_2} \frac{\dot{\boldsymbol{\sigma}}}{\|\dot{\boldsymbol{\sigma}}\|} + (c_1 + c_2 \|\boldsymbol{x}\| \\
&\quad + c_3 \left\| \boldsymbol{F}_n - \boldsymbol{x}_{1,d}^{(2)} + k_1 \|\boldsymbol{\sigma}\|^{r_1} \frac{\boldsymbol{\sigma}}{\|\boldsymbol{\sigma}\|} + k_2 \|\dot{\boldsymbol{\sigma}}\|^{r_2} \frac{\dot{\boldsymbol{\sigma}}}{\|\dot{\boldsymbol{\sigma}}\|} \right\| \\
&\quad + \eta)\frac{\boldsymbol{S}}{\|\boldsymbol{S}\|})).
\end{aligned}
\tag{5.38}
$$

Using the condition (5.29) and Eq. (5.38) can achieve that:

$$
\begin{aligned}
\dot{V}_1 &\leq \|\boldsymbol{S}\|\,(\varepsilon_1 + \varepsilon_2 \|\boldsymbol{x}\|) - \|\boldsymbol{S}\|\,(c_1 + c_2 \|\boldsymbol{x}\| + c_3 \\
&\quad \cdot \left\| \boldsymbol{F}_n - \boldsymbol{x}_{1,d}^{(2)} + k_1 \|\boldsymbol{\sigma}\|^{r_1} \frac{\boldsymbol{\sigma}}{\|\boldsymbol{\sigma}\|} + k_2 \|\dot{\boldsymbol{\sigma}}\|^{r_2} \frac{\dot{\boldsymbol{\sigma}}}{\|\dot{\boldsymbol{\sigma}}\|} \right\| + \eta) \\
&\quad + \varepsilon_3 \|\boldsymbol{S}\| \left\| \boldsymbol{F}_n - \boldsymbol{x}_{1,d}^{(2)} + k_1 \|\boldsymbol{\sigma}\|^{r_1} \frac{\boldsymbol{\sigma}}{\|\boldsymbol{\sigma}\|} + k_2 \|\dot{\boldsymbol{\sigma}}\|^{r_2} \frac{\dot{\boldsymbol{\sigma}}}{\|\dot{\boldsymbol{\sigma}}\|} \right\| \\
&\quad + \varepsilon_3 \|\boldsymbol{S}\|\,(c_1 + c_2 \|\boldsymbol{x}\| \\
&\quad + c_3 \left\| \boldsymbol{F}_n - \boldsymbol{x}_{1,d}^{(2)} + k_1 \|\boldsymbol{\sigma}\|^{r_1} \frac{\boldsymbol{\sigma}}{\|\boldsymbol{\sigma}\|} + k_2 \|\dot{\boldsymbol{\sigma}}\|^{r_2} \frac{\dot{\boldsymbol{\sigma}}}{\|\dot{\boldsymbol{\sigma}}\|} \right\| + \eta) \\
&= \|\boldsymbol{S}\|\,(\varepsilon_1 - c_1(1 - \varepsilon_3) - (1 - \varepsilon_3)\eta) \\
&\quad + \|\boldsymbol{S}\|\,\|\boldsymbol{x}\|\,(\varepsilon_2 - c_2(1 - \varepsilon_3))
\end{aligned}
\tag{5.39}
$$

$$+ \|S\| \left\| F_n - x_{1,d}^{(2)} + k_1 \|\sigma\|^{r_1} \frac{\sigma}{\|\sigma\|} + k_2 \|\dot\sigma\|^{r_2} \frac{\dot\sigma}{\|\dot\sigma\|} \right\|$$

$$\cdot (\varepsilon_3 - (1 - \varepsilon_3)c_3)$$

$$= -(1 - \varepsilon_3)\eta \|S\|$$

$$= -(1 - \varepsilon_3)\eta\sqrt{2}V_1^{\frac{1}{2}}.$$

As a result, $V_1 = 0$ when $S = 0$. It is concluded that the tracking error σ can reach the sliding manifold $S = 0$ in finite time and remain there in spite of the actuator anomalies and model uncertainties. On the sliding manifold, the equivalent dynamics can be obtained by writing $\dot{S} = 0$ as follows:

$$\ddot\sigma + k_1 \|\sigma\|^{r_1} \frac{\sigma}{\|\sigma\|} + k_2 \|\dot\sigma\|^{r_2} \frac{\dot\sigma}{\|\dot\sigma\|} = 0. \tag{5.40}$$

In the sequel, it is proved that the dynamics of Eq. (5.40) is finite-time stable. By letting $\zeta_1 = \sigma$ and $\zeta_2 = \dot\sigma$, the sliding dynamics can be represented as:

$$\begin{cases} \dot\zeta_1 = \zeta_2 \\ \dot\zeta_2 = -k_1 \dfrac{\zeta_1}{\|\zeta_1\|^{1-r_1}} - k_2 \dfrac{\zeta_2}{\|\zeta_2\|^{1-r_2}} \end{cases}. \tag{5.41}$$

Consider a Lyapunov function as:

$$V_2 = k_1 \frac{\|\zeta_1\|^{r_1+1}}{r_1+1} + \frac{\|\zeta_2\|^2}{2}. \tag{5.42}$$

The time derivative of V_2 along Eq. (5.41) is:

$$\begin{aligned} \dot{V}_2 &= k_1 \|\zeta_1\|^{r_1-1} \dot\zeta_1^T \zeta_1 - \zeta_2^T \left(k_1 \frac{\zeta_1}{\|\zeta_1\|^{1-r_1}} + k_2 \frac{\zeta_2}{\|\zeta_2\|^{1-r_2}} \right) \\ &= k_1 \|\zeta_1\|^{r_1-1} \dot\zeta_1^T \zeta_1 - k_1 \|\zeta_1\|^{r_1-1} \zeta_2^T \zeta_1 \\ &\quad - k_2 \|\zeta_2\|^{r_2-1} \zeta_2^T \zeta_2 \\ &= -k_2 \|\zeta_2\|^{r_2+1}. \end{aligned} \tag{5.43}$$

Applying LaSalle's invariance principle, the set $\{(\zeta_1, \zeta_2) : \dot{V}_2(\zeta_1, \zeta_2) = 0\}$ consists of $\zeta_2 = 0$, and the only invariant set inside $\zeta_2 = 0$ is the origin $\zeta_1 = \zeta_2 = 0$. Thus, the asymptotic convergence of ζ_1 and ζ_2 to zero is guaranteed. Further, considering the vector field Eq. (5.41) and the dilation $(1, 1, 1, \frac{1}{2-r_2}, \frac{1}{2-r_2}, \frac{1}{2-r_2})$, one can conclude that the vector field Eq. (5.41) is homogeneous of degree $\frac{r_2-1}{2-r_2} < 0$. According to Lemma 5.1, it can be concluded that system Eq. (5.40) is globally finite-time stable. Thus, the tracking error σ can be driven to the origin along $S = 0$ in finite time, although actuator faults take place. This completes the proof. $\qquad \square$

Remark 5.5. *In the sliding manifold definition Eq. (5.31) and the proposed safety control Eqs. (5.32)–(5.34), the derivatives of $\dot{\sigma}$ can be estimated on-line by the robust exact finite-time convergent differentiator [134]. The differentiator can be implemented if the higher order derivatives of the input are bounded and the finite-time escape does not exist. The differentiator transient can be driven short enough by appropriately tuning the differentiator parameters. As argued in Refs. [134, 135], the differentiator can satisfy most of feedback requirement, if the convergence of the used differentiator is adequately fast and accurate.*

Remark 5.6. *The uncertainty term, Δ_G, cannot be ignored since the limits of wind tunnel and flight tests determine a partial knowledge of HGV aerodynamic derivatives. In addition, actuator faults may be caused by the reentry thermal environment and ablation. This study with explicit consideration of both difficulties can be seen a further step of the existing literature.*

Remark 5.7. *Time-scale separation of the independent inner-loop and outer-loop designs, stemming from Eq. (5.22), is typically enforced in most of the design approaches [130, 131]. It is difficult to guarantee the finite-time stability of the overall closed-loop system. By contrast, Eq. (5.28) is integrated by Eqs. (5.13) and (5.20), which provides the basis of the proposed composite-loop design.*

Remark 5.8. *In Refs. [127, 136], multivariable TSMC design is discussed for hypersonic vehicles. The sliding manifold [127, 136] is essentially established by a decoupled treatment. Instead, based on sliding manifold of vector expression, the approach developed in this paper has twofold benefits: (1) the problem related to the decoupled design is avoided; and (2) the multivariable safety control can maintain the globally finite-time stability under actuator malfunctions and model uncertainties. These improvements have the potential to enhance the safety of operational HGVs.*

Remark 5.9. *The second term of the right hand side of Eq. (5.40) can be seen as a proportional-like control term, while the third one of the right hand side of Eq. (5.40) is considered as a differential-like control term. Eq. (5.40) can guarantee both σ and $\dot{\sigma}$ converge to zero in finite time, as detailed in the proof procedure of Theorem 5.1. The advantages of such kind of sliding manifold are:*

1) *More concise solution than a decoupled collection of single variable structures is achieved, facilitating multivariable safety control design;*

2) *Similar to the conventional terminal sliding manifold, the finite-time convergence to zero can be ensured in the sliding mode; and*

3) *As shown in the sliding manifold Eq. (5.31), the fractional power integral terms $\|\sigma\|^{r_1} \frac{\sigma}{\|\sigma\|}$ and $\|\dot{\sigma}\|^{r_2} \frac{\dot{\sigma}}{\|\dot{\sigma}\|}$ are "hidden" behind the integral action. There are no negative fractional power terms appearing in the safety*

control law Eqs. (5.32)–(5.34). In other words, the singularity problem in TSMC can be avoided completely by adopting Eq. (5.31). In this sense, the proposed approach is more straightforward as compared to those in Refs. [137, 138, 139].

5.5.2 Safety Control System Based on Adaptive Multivariable TSMC Technique

The FTC strategy against HGV actuator faults is developed, as indicated in Eqs. (5.32)–(5.34) of Theorem 5.1. However, the control parameters c_1, c_2, and c_3 may not be obtained due to the complexity of the uncertainties and actuators faults. To better address the selection of c_1, c_2, and c_3, an adaptation algorithm within the safety control scheme is proposed in Theorem 5.2.

Theorem 5.2. *Given the faulty HGV model in Eq. (5.27) and a safety control law constructed by Eqs. (5.32)–(5.34), the control parameters c_1, c_2, and c_3 can be estimated by:*

$$\dot{\hat{c}}_1 = \begin{cases} \dfrac{1}{\gamma_1} & \boldsymbol{S} \neq \boldsymbol{0} \\ 0 & \boldsymbol{S} = \boldsymbol{0} \end{cases}, \tag{5.44}$$

$$\dot{\hat{c}}_2 = \begin{cases} \dfrac{1}{\gamma_2}\|\boldsymbol{x}\| & \boldsymbol{S} \neq \boldsymbol{0} \\ 0 & \boldsymbol{S} = \boldsymbol{0} \end{cases}, \tag{5.45}$$

$$\dot{\hat{c}}_3 = \begin{cases} \dfrac{1}{\gamma_3}\left\| \boldsymbol{F}_n - \boldsymbol{x}_{1,d}^{(2)} + k_1\|\boldsymbol{\sigma}\|^{r_1}\dfrac{\boldsymbol{\sigma}}{\|\boldsymbol{\sigma}\|} + k_2\|\boldsymbol{\sigma}\|^{r_2}\dfrac{\dot{\boldsymbol{\sigma}}}{\|\dot{\boldsymbol{\sigma}}\|}\right\| & \boldsymbol{S} \neq \boldsymbol{0} \\ 0 & \boldsymbol{S} = \boldsymbol{0} \end{cases}. \tag{5.46}$$

Note that γ_i are positive design parameters, $\hat{c}_i(0) > 0$, and $i = 1, 2, 3$.

Proof. : The parameter errors are defined as:

$$\tilde{c}_1 = \hat{c}_1 - \frac{\varepsilon_1}{1 - \varepsilon_3}, \tilde{c}_2 = \hat{c}_2 - \frac{\varepsilon_2}{1 - \varepsilon_3}, \tilde{c}_3 = \hat{c}_3 - \frac{\varepsilon_3}{1 - \varepsilon_3}. \tag{5.47}$$

A Lyapunov function is selected as:

$$V_3 = \|\boldsymbol{S}\| + \frac{1 - \varepsilon_2}{2}\gamma_1\tilde{c}_1^2 + \frac{1 - \varepsilon_2}{2}\gamma_2\tilde{c}_2^2 + \frac{1 - \varepsilon_2}{2}\gamma_3\tilde{c}_3^2. \tag{5.48}$$

Noting that $\dot{\tilde{c}}_i = \dot{\hat{c}}_i$, $i = 1, 2, 3$, one can obtain the time derivative of V_3 along the trajectories of Eq. (5.27) when $\boldsymbol{S} \neq \boldsymbol{0}$

$$\dot{V}_3 = \frac{\boldsymbol{S}^T\dot{\boldsymbol{S}}}{\|\boldsymbol{S}\|} + (1 - \varepsilon_3)\gamma_1\tilde{c}_1\dot{\hat{c}}_1 + (1 - \varepsilon_3)\gamma_2\tilde{c}_2\dot{\hat{c}}_2$$

$$+ (1 - \varepsilon_3)\gamma_3\tilde{c}_3\dot{\hat{c}}_3$$

$$= \frac{S^T}{\|S\|}(\dot{f}_1 + \Delta_F + (G_n + \Delta_G)\rho - (\hat{c}_1 + \hat{c}_2 \|x\|$$

$$+ \hat{c}_3 \left\| F_n - x_{1,d}^{(2)} + k_1 \|\sigma\|^{r_1} \frac{\sigma}{\|\sigma\|} \right.$$

$$+ k_2 \|\dot{\sigma}\|^{r_2} \frac{\dot{\sigma}}{\|\dot{\sigma}\|} \left\| + \eta) \frac{S}{\|S\|} - (\Delta_G \Lambda + G_n(\Lambda - I))\right.$$

$$\cdot G_n^{-1} \left[F_n - x_{1,d}^{(2)} + k_1 \|\sigma\|^{r_1} \frac{\sigma}{\|\sigma\|} + k_2 \|\dot{\sigma}\|^{r_2} \frac{\dot{\sigma}}{\|\dot{\sigma}\|} \right] \qquad (5.49)$$

$$- (\Delta_G \Lambda + G_n(\Lambda - I))G_n^{-1}(\hat{c}_1 + \hat{c}_2 \|x\|$$

$$+ \hat{c}_3 \left\| F_n - x_{1,d}^{(2)} + k_1 \|\sigma\|^{r_1} \frac{\sigma}{\|\sigma\|} + k_2 \|\dot{\sigma}\|^{r_2} \frac{\dot{\sigma}}{\|\dot{\sigma}\|} \right\|$$

$$+ \eta) \frac{S}{\|S\|}) + (1 - \varepsilon_3)\hat{c}_1 - \varepsilon_1 + (1 - \varepsilon_3)\hat{c}_2 \|x\|$$

$$- \varepsilon_2 \|x\| + (1 - \varepsilon_3)\hat{c}_3$$

$$\cdot \left\| F_n - x_{1,d}^{(2)} + k_1 \|\sigma\|^{r_1} \frac{\sigma}{\|\sigma\|} + k_2 \|\dot{\sigma}\|^{r_2} \frac{\dot{\sigma}}{\|\dot{\sigma}\|} \right\|$$

$$- \varepsilon_3 \left\| F_n - x_{1,d}^{(2)} + k_1 \|\sigma\|^{r_1} \frac{\sigma}{\|\sigma\|} + k_2 \|\dot{\sigma}\|^{r_2} \frac{\dot{\sigma}}{\|\dot{\sigma}\|} \right\|.$$

Applying the condition Eq. (5.29) to Eq. (5.49) gives:

$$\dot{V}_3 \leq \left\| \dot{f}_1 + \Delta_F + (G_n + \Delta_G)\rho \right\|$$

$$+ \left\| (\Delta_G \Lambda + G_n(\Lambda - I))G_n^{-1} \right\|$$

$$\cdot \left\| F_n - x_{1,d}^{(2)} + k_1 \|\sigma\|^{r_1} \frac{\sigma}{\|\sigma\|} + k_2 \|\dot{\sigma}\|^{r_1} \frac{\dot{\sigma}}{\|\dot{\sigma}\|} \right\|$$

$$- (\hat{c}_1 + \hat{c}_2 \|x\| + \hat{c}_3 \left\| F_n - x_{1,d}^{(2)} + k_1 \|\sigma\|^{r_1} \frac{\sigma}{\|\sigma\|} \right.$$

$$+ k_2 \|\dot{\sigma}\|^{r_1} \frac{\dot{\sigma}}{\|\dot{\sigma}\|} \left\| + \eta) + \left\| (\Delta_G \Lambda + G_n(\Lambda - I))G_n^{-1} \right\|\right.$$

$$\cdot (\hat{c}_1 + \hat{c}_2 \|x\| + \hat{c}_3 \left\| F_n - x_{1,d}^{(2)} + k_1 \|\sigma\|^{r_1} \frac{\sigma}{\|\sigma\|} \right.$$

$$+ k_2 \|\dot{\sigma}\|^{r_1} \frac{\dot{\sigma}}{\|\dot{\sigma}\|} \left\| + \eta) + (1 - \varepsilon_3)\hat{c}_1 - \varepsilon_1 \right.$$

$$+ (1 - \varepsilon_3)\hat{c}_2 \|x\| - \varepsilon_2 \|x\| + (1 - \varepsilon_3)\hat{c}_3$$

$$\cdot \left\| \boldsymbol{F}_n - x_{1,d}^{(2)} + k_1 \|\boldsymbol{\sigma}\|^{r_1} \frac{\boldsymbol{\sigma}}{\|\boldsymbol{\sigma}\|} + k_2 \|\dot{\boldsymbol{\sigma}}\|^{r_2} \frac{\dot{\boldsymbol{\sigma}}}{\|\dot{\boldsymbol{\sigma}}\|} \right\|$$

$$- \varepsilon_3 \left\| \boldsymbol{F}_n - x_{1,d}^{(2)} + k_1 \|\boldsymbol{\sigma}\|^{r_1} \frac{\boldsymbol{\sigma}}{\|\boldsymbol{\sigma}\|} + k_2 \|\dot{\boldsymbol{\sigma}}\|^{r_2} \frac{\dot{\boldsymbol{\sigma}}}{\|\dot{\boldsymbol{\sigma}}\|} \right\| \qquad (5.50)$$

$$\leq \varepsilon_1 + \varepsilon_2 \|\boldsymbol{x}\| - (\hat{c}_1 + \hat{c}_2 \|\boldsymbol{x}\| + \hat{c}_3$$

$$\cdot \left\| \boldsymbol{F}_n - x_{1,d}^{(2)} + k_1 \|\boldsymbol{\sigma}\|^{r_1} \frac{\boldsymbol{\sigma}}{\|\boldsymbol{\sigma}\|} + k_2 \|\dot{\boldsymbol{\sigma}}\|^{r_2} \frac{\dot{\boldsymbol{\sigma}}}{\|\dot{\boldsymbol{\sigma}}\|} \right\| + \eta)$$

$$+ \varepsilon_3 \left\| \boldsymbol{F}_n - x_{1,d}^{(2)} + k_1 \|\boldsymbol{\sigma}\|^{r_1} \frac{\boldsymbol{\sigma}}{\|\boldsymbol{\sigma}\|} + k_2 \|\dot{\boldsymbol{\sigma}}\|^{r_2} \frac{\dot{\boldsymbol{\sigma}}}{\|\dot{\boldsymbol{\sigma}}\|} \right\|$$

$$+ \varepsilon_3 (\hat{c}_1 + \hat{c}_2 \|\boldsymbol{x}\| + \hat{c}_3 \left\| \boldsymbol{F}_n - x_{1,d}^{(2)} + k_1 \|\boldsymbol{\sigma}\|^{r_1} \frac{\boldsymbol{\sigma}}{\|\boldsymbol{\sigma}\|} \right.$$

$$\left. + k_2 \|\dot{\boldsymbol{\sigma}}\|^{r_2} \frac{\dot{\boldsymbol{\sigma}}}{\|\dot{\boldsymbol{\sigma}}\|} \right\| + \eta) + (1 - \varepsilon_3)\hat{c}_1 - \varepsilon_1$$

$$+ (1 - \varepsilon_3)\hat{c}_2 \|\boldsymbol{x}\| - \varepsilon_2 \|\boldsymbol{x}\| + (1 - \varepsilon_3)\hat{c}_3$$

$$\cdot \left\| \boldsymbol{F}_n - x_{1,d}^{(2)} + k_1 \|\boldsymbol{\sigma}\|^{r_1} \frac{\boldsymbol{\sigma}}{\|\boldsymbol{\sigma}\|} + k_2 \|\dot{\boldsymbol{\sigma}}\|^{r_2} \frac{\dot{\boldsymbol{\sigma}}}{\|\dot{\boldsymbol{\sigma}}\|} \right\|$$

$$- \varepsilon_3 \left\| \boldsymbol{F}_n - x_{1,d}^{(2)} + k_1 \|\boldsymbol{\sigma}\|^{r_1} \frac{\boldsymbol{\sigma}}{\|\boldsymbol{\sigma}\|} + k_2 \|\dot{\boldsymbol{\sigma}}\|^{r_2} \frac{\dot{\boldsymbol{\sigma}}}{\|\dot{\boldsymbol{\sigma}}\|} \right\|$$

$$= -(1 - \varepsilon_3)\eta < 0.$$

The condition Eq. (5.50) implies that the values of V_3 and $\boldsymbol{S}$ will approach to zero in finite time t_f, i.e., $V_3(t_f) = 0$, and it can be verified that $t_f \leq t_0 + \frac{V_3(t_0)}{\eta(1-\varepsilon_3)}$. Since the value of V_3 is bounded, $\tilde{c}_i$ (and hence $\hat{c}_i$) are all bounded. Moreover, in accordance with Eq. (5.45), the solution of $\hat{c}_2$ is:

$$\hat{c}_2(t) = \frac{1}{\gamma_2} \int_{t_0}^{t_f} \|\boldsymbol{x}(\tau)\| \, d\tau + \hat{c}_2(0), \quad \boldsymbol{S} \neq \boldsymbol{0}. \qquad (5.51)$$

When $\hat{c}_2$ is bounded and the integral is nonnegative, the state variable $\boldsymbol{x}$ of Eq. (5.27) is bounded for $t_0 < t < t_f$. □

Remark 5.10. *$\boldsymbol{u}_b$ is considered as nominal control, while $\boldsymbol{u}_d$ is designed to compensate for the detrimental impact of HGV actuator faults and model uncertainties. The parameters c_1, c_2, and c_3 in Eq. (5.33) are linked to the HGV actuator faults. It is worth to emphasize that the values of c_1, c_2, and c_3 are tuned in response to the knowledge of the current status of the HGV.*

Remark 5.11. *It is noted that $\boldsymbol{u}_d$ contains a switching term $\frac{\boldsymbol{S}}{\|\boldsymbol{S}\|}$. Due to non-linearities, noisy measurements, and nonideal switching, the control chattering* *exists. For alleviating the chattering phenomenon, one option is to replace the switching function $\frac{S}{\|S\|}$ of Eq. (5.34) as:*

$$\text{sat}(S, \Phi) = \begin{cases} \dfrac{S}{\Phi} & \|S\| \leq \Phi \\ \dfrac{S}{\|S\|} & \|S\| > \Phi \end{cases}, \tag{5.52}$$

where Φ is a small positive constant. Here, Φ helps define a boundary layer about the sliding manifold $S = 0$ inside which an acceptably close approximation to ideal sliding takes place. Provided the states evolve with time inside the boundary layer, no adaptation of the switching gains takes place. If a fault occurs that starts to make the sliding motion degrade so that the states evolve outside the boundary, i.e., $\|S\| > \Phi$, then the gains $\hat{c}_i$ increase in magnitude to force the states back into the boundary layer around the sliding manifold. The adaptive tuning laws Eqs. (5.44)–(5.46) are modified as [140, 141]:

$$\dot{\hat{c}}_1 = \begin{cases} \dfrac{1}{\gamma_1} & \|S\| > \Phi \\ 0 & \|S\| \leq \Phi \end{cases}, \tag{5.53}$$

$$\dot{\hat{c}}_2 = \begin{cases} \dfrac{1}{\gamma_2}\|x\| & \|S\| > \Phi \\ 0 & \|S\| \leq \Phi \end{cases}, \tag{5.54}$$

$$\dot{\hat{c}}_3 = \begin{cases} \dfrac{1}{\gamma_3} \left\| F_n - x_{1,d}^{(2)} + k_1 \|\sigma\|^{r_1} \dfrac{\sigma}{\|\sigma\|} + k_2 \|\dot{\sigma}\|^{r_2} \dfrac{\dot{\sigma}}{\|\dot{\sigma}\|} \right\| & \|S\| > \Phi, \\ 0 & \|S\| \leq \Phi \end{cases}. \tag{5.55}$$

The designer has to tradeoff between tracking accuracy and chattering phenomenon when adjusting the parameter Φ. A time-varying dead-zone modification is proposed in Ref. [142]. The resulting method can improve the adaptive law robustness to measurement or system noises.

5.6 Simulation Results

5.6.1 HGV Flight Condition and Simulation Scenarios

The initial flight conditions of the HGV are: $V(0) = 3000\,\text{m/s}$, $H = 30\,000\,\text{m}$, $\mu(0) = 2\,°$, $\alpha(0) = 2\,°$, $\beta(0) = 2\,°$, and $p(0) = q(0) = r(0) = 0$. The model uncertainties, actuator faults, and sensed signals with white noises are introduced to assess the performance of the developed safety control scheme.

1) The uncertainties corresponding to the roll, pitch, and yaw moments of inertia (I_{xx}, I_{yy}, I_{zz}) are 10% of the nominal values. The maximal 20% mismatch exists in the HGV mass, C_l, C_m, and C_n, respectively.

2) Focusing on the fault pattern of the HGV actuators, the gain faults and bias faults are included with consideration of both time-invariant and time-varying cases, as can be found in Eqs. (5.56)–(5.57), respectively.

Actuator time-invariant faults:

$$\lambda_1 = \begin{cases} 1 & 0 \leq t < 9 \\ 0.75 & t > 9 \end{cases}, \rho_1 = \begin{cases} 0 & 0 \leq t < 9 \\ 2 & t > 9 \end{cases},$$

$$\lambda_2 = \begin{cases} 1 & 0 \leq t < 9 \\ 0.75 & t > 9 \end{cases}, \rho_2 = \begin{cases} 0 & 0 \leq t < 9 \\ -3 & t > 9 \end{cases}, \quad (5.56)$$

$$\lambda_3 = \begin{cases} 1 & 0 \leq t < 9 \\ 0.75 & t > 9 \end{cases}, \rho_3 = \begin{cases} 0 & 0 \leq t < 9 \\ 2 & t > 9 \end{cases}.$$

Actuator time-varying faults:

$$\lambda_1 = \begin{cases} 1 & 0 \leq t < 9 \\ 0.75 - \dfrac{t-9}{60} & 9 \leq t \leq 15 \\ 0.65 & t > 15 \end{cases},$$

$$\rho_1 = \begin{cases} 0 & 0 \leq t < 9 \\ 2 + \dfrac{t-9}{3} & 9 \leq t \leq 15 \\ 4 & t > 15 \end{cases},$$

$$\lambda_2 = \begin{cases} 1 & 0 \leq t < 9 \\ 0.75 - \dfrac{t-9}{30} & 9 \leq t \leq 15 \\ 0.55 & t > 15 \end{cases},$$

$$\rho_2 = \begin{cases} 0 & 0 \leq t < 9 \\ -3 - \dfrac{t-9}{6} & 9 \leq t \leq 15 \\ -4 & t > 15 \end{cases},$$

$$(5.57)$$

$$\lambda_3 = \begin{cases} 1 & 0 \leq t < 9 \\ 0.7 - \dfrac{t-9}{30} & 9 \leq t \leq 15 \\ 0.5 & t > 15 \end{cases},$$

$$\rho_3 = \begin{cases} 0 & 0 \leq t < 9 \\ 2 + \dfrac{t-9}{3} & 9 \leq t \leq 15 \\ 4 & t > 15 \end{cases}.$$

3) The white noise with a mean of 0 and covariance of 0.01 is injected into each measurement channel.

Two scenarios are studied to demonstrate the use of the algorithms for HGV attitude tracking control. (1) *Scenario I*: the model uncertainty, time-invariant actuator gain and bias faults, and measurement noises are considered; and (2) *Scenario II*: the model uncertainty, time-varying actuator gain and bias faults, and measurement noises are involved.

The control parameters are selected as: $k_1 = 3$, $k_2 = 4$, $r_1 = \frac{1}{3}$, $r_2 = \frac{1}{2}$, $\eta = 0.1$, and $\Phi = 0.15$. In the adaptive laws: $\gamma_1 = \frac{1}{40}$, $\gamma_2 = \frac{1}{15}$, and $\gamma_3 = \frac{1}{10}$. $\hat{c}_1(0) = \hat{c}_2(0) = \hat{c}_3(0) = 0$.

For quantitatively evaluating the tracking performance, an index is defined as:

$$\sigma_{p,j} = \sqrt{\frac{1}{t_2 - t_1} \int_{t_1}^{t_2} |\sigma_j|^2 \, d\tau}, \quad j = \mu, \alpha, \beta, \tag{5.58}$$

where $[t_1, t_2]$ covers the time frame of the overall simulation. The defined metric is the scalar valued L_2 norm, as a measure of average tracking performance.

5.6.2 Simulation Analysis of Scenario I

It is highlighted in Fig. 5.2(a) that after the actuator faults take place $(t \geq 9\,\text{s})$, the reference signal can be quickly tracked under the proposed safety control scheme. As can be seen from Figs. 5.2(b) and 5.2(c), the AOA and sideslip angle can converge to the intended values within finite time in the presence or absence of actuator faults. Focusing on Fig. 5.2, the developed safety control scheme allows the HGV to follow the prescribed tracking profiles as closely as possible under the actuator faults and model uncertainties. The defined indices in Eq. (5.58) with respect to the bank angle, AOA, and sideslip angle are $0.4068\,°$, $0.1904\,°$, and $0.1764\,°$, respectively. Based on Fig. 5.3, the amplitude of the actuators becomes larger than that of the normal case, such that the effects induced by the faults can be eliminated. Key observations from Fig. 5.4 are : (1) the estimated values of the parameters ($\hat{c}_1$, $\hat{c}_2$, and $\hat{c}_3$) hold at constant values to counteract model uncertainties ($0 \leq t < 9\,\text{s}$) and (2) the estimated values respond appropriately by applying the adaptive laws after the occurrence of the actuator faults.

5.6.3 Simulation Analysis of Scenario II

The performance against actuator time-varying faults is evaluated in Scenario II. From Fig. 5.5, the tracking performance is satisfactory, when the actuator time-varying malfunctions, measurement noises, and model uncertainties simultaneously exist. The HGV states can be steered to the intended values in a timely manner. The defined metrics corresponding to the bank angle, AOA, and sideslip angle are $0.4671\,°$, $0.2025\,°$, and $0.2137\,°$, respectively. As compared to those in Scenario I (see Table 5.1), the performance is

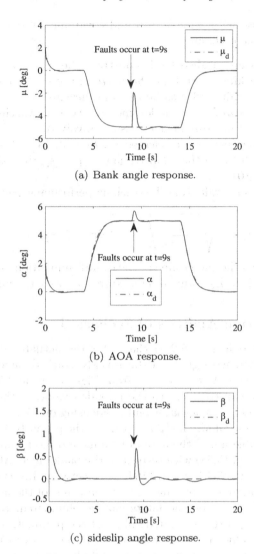

(a) Bank angle response.

(b) AOA response.

(c) sideslip angle response.

FIGURE 5.2: The curves of the tracking angles in Scenario I.

decreased by 14.82%, 6.36%, 21.15%, respectively. This condition arises due to that the impact of time-varying faults is worse than that of time-invariant ones. The deflections of the actuators and the adaptation process of the control gains are depicted in Figs. 5.6 and 5.7, respectively. The control gains can be promptly updated in response to the time-varying faults. The actuators are appropriately managed to maintain the HGV safety. In summary, the applicability of the developed safety control scheme is further verified through the simulation studies of Scenario II.

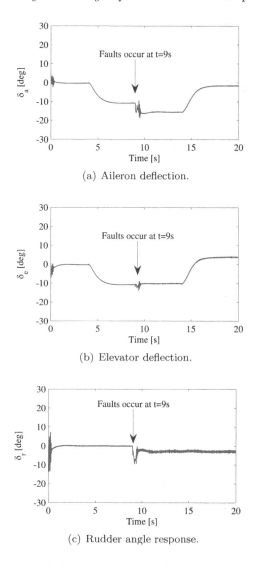

(a) Aileron deflection.

(b) Elevator deflection.

(c) Rudder angle response.

FIGURE 5.3: The curves of the deflections in Scenario I.

5.7   Concluding Remarks

A safety control architecture, including the multivariable integral TSMC and adaptive approaches suitable for HGV attitude tracking control, is developed against actuator faults and model uncertainties. The unique advantages of the proposed method lie in three aspects.

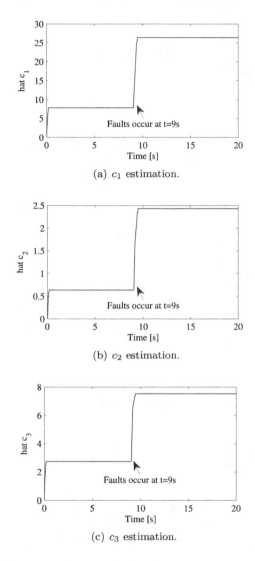

(a) c_1 estimation.

(b) c_2 estimation.

(c) c_3 estimation.

FIGURE 5.4: Curves of the adaptive gains in Scenario I.

1) The finite-time stability of the faulty HGV can be guaranteed so that unacceptable HGV behaviors are not created by actuator gain and bias malfunctions;

2) The composite-loop design under actuator faults is achieved on the basis of control-oriented model, without the need of the time-scale separation principle; and

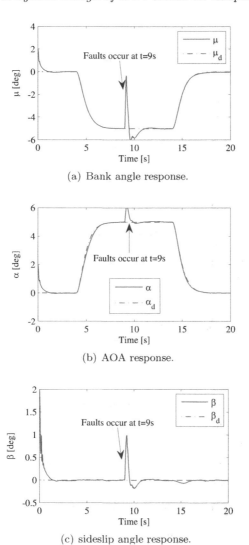

(a) Bank angle response.

(b) AOA response.

(c) sideslip angle response.

FIGURE 5.5: The curves of the tracking angles in Scenario II.

TABLE 5.1: Performance index.

-	$\sigma_{p,\mu}$	$\sigma_{p,\alpha}$	$\sigma_{p,\beta}$
Scenario I	0.4068	0.1904	0.1764
Scenario II	0.4671	0.2025	0.2137

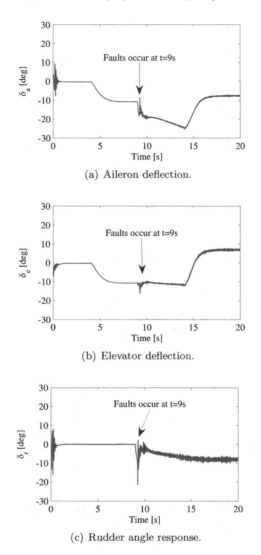

(a) Aileron deflection.

(b) Elevator deflection.

(c) Rudder angle response.

FIGURE 5.6: The curves of the deflections in Scenario II.

3) The multivariable integral TSMC method is presented to enable integration into the HGV safety control design, instead of the decoupled single-input and single-output method.

The simulations of a full nonlinear model of the HGV dynamics show that the investigated scheme can be successfully employed to handle scenarios involving actuator faults and model uncertainties.

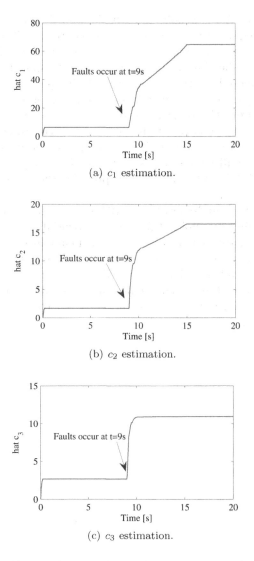

(a) c_1 estimation.

(b) c_2 estimation.

(c) c_3 estimation.

FIGURE 5.7: Curves of the adaptive gains in Scenario II.

5.8   Notes

This chapter presents a safety control strategy for a hypersonic gliding vehicle (HGV) subject to actuator malfunctions and model uncertainties. The control-oriented model of the HGV is established according to the HGV kinematic and aerodynamic models. A composite-loop design for HGV safety control under actuator faults is subsequently developed, where newly developed multivariable integral terminal sliding-mode control (TSMC) and adaptive techniques are integrated. The simulations show that the HGV can handle the time-invariant, the time-varying actuator faults and model uncertainties well with the proposed safety control design techniques.

Time-scale separation of the independent inner-loop and outer-loop designs, stemming from Eq. (5.22), is typically enforced in most of the design approaches [130, 131]. One needs to develop two controllers corresponding to the separated loops. However, it is difficult to guarantee the finite-time stability of the overall closed-loop system. In this study, Eq. (5.28) is integrated by Eqs. (5.13) and (5.20), which provides the basis of the proposed composite-loop design.

In Refs. [127, 136], multivariable TSMC design is discussed for hypersonic vehicles. The sliding manifold [127, 136] is essentially established by a decoupled treatment. Instead, based on sliding manifold of vector expression, the approach developed in this chapter has twofold benefits: (1) the problem related to the decoupled design is avoided; and (2) the multivariable safety control can maintain the globally finite-time stability under actuator malfunctions. These improvements have the potential to enhance the safety of operational HGVs.

Chapter 6

Safety Control System Design of HGV Based on Fixed-Time Observer

6.1 Introduction

The increasing complexity and automation render the HGVs susceptible to component/system faults. Substantial performance degradation and even catastrophic consequences can be attributed to in-flight failures. Although various degrees of success in HGV safety control design have been achieved, there still exist several problems to be further investigated. 1) From a safety point of view, it is highly desirable that the fault diagnosis and the corresponding accommodation can be completed in a timely manner [4, 129]. So, how to adopt finite/fixed-time stability in the HGV fault diagnosis and then accommodation requires extensive research. 2) Due to short of wind tunnel facilities and flight experiments, a severe difficulty in designing HGV control systems results from the large uncertainties and perturbations inherent to the HGV model [143]. Not only uncertainty exists in dynamic matrix, but also in control input matrix. This factor renders a great challenge of safety control design. 3) Sliding mode observer (SMO) and sliding mode control (SMC) start being exploited for HGV safety control system design. Nevertheless, in most of the SMO and SMC approaches proposed to date, a multi-input problem with m inputs is recast as a decoupled problem involving m single-input structures. Multivariable design of SMO and SMC is demanded rather than the decoupled treatment, by considering strong couplings and nonlinearity inherent to an HGV.

Motivated by the aforementioned difficulties, particular attention is devoted to achieving fixed-time fault estimation and finite-time fault accommodation within the context of multivariable design. Thus, the developed safety control scheme can provide accommodation for HGV actuator faults and model uncertainties. The major contributions are stated as follows.

1. When comparing to the finite-time SMO [136, 144, 145] and the sliding mode based disturbance observer [146, 147, 148, 149], the proposed fixed-time SMO can ensure that the estimation error of the "lumped disturbance" converges to a small vicinity of zero in fixed time. Moreover, the convergence time of the designed disturbance observer is indepen-

117

dent of initial conditions, while elegant solutions can be achieved by vector form design.

2. The safety control presented in this chapter is integrated by fixed-time SMO and finite-time control. The resulting safety control can promptly enforce the trajectory of the faulty HGV converging to a small vicinity of origin, without the need of excessive control efforts. It should be mentioned that both model uncertainties and actuator faults are explicitly considered over the design phase. By resorting to the proposed safety control scheme, corrective reactions can be taken in response to the actuator faults for fulfilling the stringent requirement of HGV safety. To the best of the authors' knowledge, there exist few papers focusing on this aspect.

3. In opposition to the multivariable SMC design for a hypersonic vehicle in [136], the developed approach of this study can avoid the problem associated with the decoupled design and ensure the globally finite-time stability in spite of actuator faults. This study is applicable especially in the case where the strong couplings are exposed on HGV aerodynamics. The use of multivariable design avoids the necessity for the decoupled design with m single-input and single-output (SISO) structures. These improvements have the potential of enhancing the safety of operational HGVs, since the coupling and inherent functional redundancy of an HGV have been better exploited in such a multivariable design approach.

The rest of this chapter is arranged as follows. The control-oriented HGV model, actuator fault mode, and problem statement are given in Chapter 6.2. An HGV safety control scheme is proposed against actuator malfunctions and model uncertainties in Chapter 6.3, where the fixed-time SMO and the finite-time SMC based safety control are presented with the aid of multivariable design. In Chapter 6.4, the performance of the developed safety control is evaluated by means of simulations of a full nonlinear HGV model. Chapter 6.5 includes a discussion of the conclusions.

6.2 HGV Modeling and Problem Statement

6.2.1 HGV Dynamics

The HGV is modeled based on the assumption of a rigid vehicle structure, a flat, non-rotating Earth and uniform gravitational field. Therefore, the inertial position coordinates are written as:

$$\begin{cases} \dot{x} = V \cos\gamma \cos\chi \\ \dot{y} = V \cos\gamma \sin\chi \ , \\ \dot{z} = -V \sin\gamma \end{cases} \tag{6.1}$$

where x, y, and z stand for the positions with respect to x, y, and z directions of the Earth-fixed reference frame, V specifies the total velocity of the HGV, χ and μ represent the flight-path angle and the bank angle, respectively.

The force equations are expressed as:

$$\begin{cases} \dot{V} = -g\sin\gamma - \dfrac{QS_rC_D}{m} \\ \dot{\chi} = \dfrac{QS_r}{mV\cos\gamma}(C_L\sin\mu + C_Y\cos\mu) \\ \dot{\gamma} = -\dfrac{g}{V}\cos\gamma + \dfrac{QS_r}{mV}(C_L\cos\mu - C_Y\sin\mu) \end{cases} \tag{6.2}$$

where g, Q, S_r, m and γ denote the gravitational constant, the dynamic pressure, the reference area, the mass of the HGV, and the heading angle, C_L, C_D, and C_Y are the aerodynamic coefficients with respect to lift, drag, and side force, respectively.

The kinematic model of attitude is described as:

$$\begin{cases} \dot{\mu} = \sec\beta(p\cos\alpha + r\sin\alpha) \\ \quad + \dfrac{QS_rC_L}{mV}(\tan\gamma\sin\mu + \tan\beta) \\ \quad + \dfrac{QS_rC_Y}{mV}\tan\gamma\cos\mu - \dfrac{g}{V}\cos\gamma\cos\mu\tan\beta \\ \dot{\alpha} = q - \tan\beta(p\cos\alpha + r\sin\alpha) \\ \quad + \dfrac{1}{mV\cos\beta}(mg\cos\gamma\cos\mu - QS_rC_L) \\ \dot{\beta} = -r\cos\alpha + p\sin\alpha \\ \quad + \dfrac{1}{mV}(QS_rC_Y + mg\cos\gamma\sin\mu) \end{cases} \tag{6.3}$$

where α and β denote the angle of attack (AOA) and the sideslip angle, respectively.

The dynamic model of attitude is given as:

$$\begin{cases} \dot{p} = \dfrac{QS_r\bar{b}C_l}{I_{xx}} \\ \dot{q} = \dfrac{QS_r\bar{b}C_m + (I_{zz} - I_{xx})pr}{I_{yy}} \ , \\ \dot{r} = \dfrac{QS_r\bar{b}C_n + (I_{xx} - I_{yy})pq}{I_{zz}} \end{cases} \tag{6.4}$$

where $\bar{b}$ denotes the wing span of the HGV, C_l, C_m, and C_n represent the moment coefficients of the rolling, pitching, and yawing channels, I_{xx}, I_{yy}, and I_{zz} denote moments of inertia with respect to x, y, and z coordinate, respectively.

The aerodynamic force and moment coefficients are:

$$C_D = C_D(M, \alpha)$$

$$C_Y = C_{Y,\beta}(M, \alpha)\beta + C_{Y,\delta_r}(M, \alpha)\delta_r$$

$$C_L = C_L(M, \alpha) + C_{L,\delta_e}(M, \alpha)\delta_e$$

$$C_l = C_{l,\beta}(M, \alpha)\beta + C_{l,\delta_a}(M, \alpha)\delta_a + C_{l,p}(M, \alpha)\frac{p\bar{b}}{2V} \quad , \tag{6.5}$$

$$C_m = C_{m,\alpha}(M, \alpha) + C_{m,\delta_e}(M, \alpha)\delta_e + C_{m,q}(M, \alpha)\frac{q\bar{b}}{2V}$$

$$C_n = C_{n,\beta}(M, \alpha)\beta + C_{n,\delta_r}(M, \alpha)\delta_r + C_{n,r}(M, \alpha)\frac{r\bar{b}}{2V}$$

where δ_a, δ_e, and δ_r denote the control deflections of the aileron, elevator, and rudder, respectively. Note that $C_{Y,\beta}$ and C_{Y,δ_r} are the partial derivatives of C_Y with respect to β and δ_r, respectively. C_{L,δ_e} stands for the partial derivative of C_L with respect to δ_e. $C_{l,\beta}$, C_{l,δ_a}, and $C_{l,p}$ are the partial derivatives of C_l with respect to β, δ_a, and p, respectively. $C_{m,\alpha}$, C_{m,δ_e}, and $C_{m,q}$ are the partial derivatives of C_m with respect to α, δ_e, and q, respectively. $C_{n,\beta}$, C_{n,δ_r}, and $C_{n,r}$ are the partial derivatives of C_n with respect to β, δ_r, and r, respectively.

6.2.2 Control-Oriented Model Subject to Actuator Faults and Uncertainties

The control-oriented model of the HGV is established by combining the kinematic model and the dynamic model of the HGV attitude, based on which the so-called composite-loop safety control design can be achieved.

As far as an attitude control system is concerned, μ, α, and β can be gathered into a vector $x_1 = [\mu, \alpha, \beta]^T$. In terms of Eq. (6.3), one can obtain:

$$\begin{cases} \dot{\mu} = \sec\beta(p\cos\alpha + r\sin\alpha) + f_\mu \\ \dot{\alpha} = q - \tan\beta(p\cos\alpha + r\sin\alpha) + f_\alpha \\ \dot{\beta} = -r\cos\alpha + p\sin\alpha + f_\beta \end{cases} \tag{6.6}$$

where

$$\begin{cases} f_\mu = \dfrac{QSC_L}{mV}(\tan\gamma\sin\mu + \tan\beta) \\ \qquad + \dfrac{QSC_Y}{mV}\tan\gamma\cos\mu - \dfrac{g}{V}\cos\gamma\cos\mu\tan\beta \\ f_\alpha = \dfrac{1}{mV\cos\beta}(mg\cos\gamma\cos\mu - QSC_L) \\ f_\beta = \dfrac{1}{mV}(QSC_Y\cos\beta + mg\cos\gamma\sin\mu) \end{cases} \tag{6.7}$$

By defining $x_2 = [p, q, r]^T$, Eqs. (6.6)-(6.7) can be described as:

$$\dot{x}_1 = f_1 + g_1 x_2, \tag{6.8}$$

where $f_1 = [f_\mu, f_\alpha, f_\beta]^T$ and

$$g_1 = \begin{bmatrix} \sec\beta\cos\alpha & 0 & \sec\beta\sin\alpha \\ -\tan\beta\cos\alpha & 1 & -\tan\beta\sin\alpha \\ \sin\alpha & 0 & -\cos\alpha \end{bmatrix}. \tag{6.9}$$

By accounting for Eqs. (6.4)-(6.5), one can render:

$$\begin{cases} \dot{p} = f_p + \dfrac{QS_r\bar{b}C_{l,\delta_a}}{I_{xx}}\delta_a \\[2mm] \dot{q} = f_q + \dfrac{QS_r\bar{b}C_{m,\delta_e}}{I_{yy}}\delta_e \\[2mm] \dot{r} = f_r + \dfrac{QS_r\bar{b}C_{n,\delta_r}}{I_{zz}}\delta_r \end{cases}, \tag{6.10}$$

where

$$\begin{cases} f_p = \dfrac{QS_r\bar{b}(C_{l,\beta}\beta + C_{l,p}\frac{p\bar{b}}{2V})}{I_{xx}} \\[3mm] f_q = \dfrac{QS_r\bar{b}(C_{m,\alpha} + C_{m,q}\frac{q\bar{b}}{2V}) + (I_{zz} - I_{xx})pr}{I_{yy}} \\[3mm] f_r = \dfrac{QS_r\bar{b}(C_{n,\beta}\beta + C_{n,r}\frac{r\bar{b}}{2V}) + (I_{xx} - I_{yy})pq}{I_{zz}} \end{cases}. \tag{6.11}$$

Further, Eqs. (6.10)-(6.11) can be formed as:

$$\dot{x}_2 = f_2 + g_2 u, \tag{6.12}$$

where $f_2 = [f_p, f_q, f_r]^T$, $u = [\delta_a, \delta_e, \delta_r]^T$, and

$$g_2 = \begin{bmatrix} \dfrac{QS_r\bar{b}C_{l,\delta_a}}{I_{xx}} & 0 & 0 \\[3mm] 0 & \dfrac{QS_r\bar{b}C_{m,\delta_e}}{I_{yy}} & 0 \\[3mm] 0 & 0 & \dfrac{QS_r\bar{b}C_{n,\delta_r}}{I_{zz}} \end{bmatrix}. \tag{6.13}$$

Gain fault and bias fault are the faults commonly occurring on flight actuators. In this work, the actuator fault model including both sorts of faults is generally formed as:

$$u_F = \Lambda u + \rho, \tag{6.14}$$

where $\Lambda = \text{diag}\{\lambda_1, \lambda_2, \lambda_3\}$ represents the gain fault and $\rho = [\rho_a, \rho_e, \rho_r]^T$

denotes the bias fault, respectively. Note that $0 < \lambda_i \leq 1$, $i = 1, 2, 3$. Thus, in the presence of actuator faults, Eq. (6.12) is represented as:

$$
\begin{aligned}
\dot{x}_2 &= f_2 + g_2(\Lambda u + \rho) \\
&= f_2 + g_2 u + g_2(\Lambda - I)u + g_2\rho.
\end{aligned} \tag{6.15}
$$

Note that u in Eqs. (6.12), (6.14), and (6.15) represents the control input vector under normal conditions, while u_F describes the control input vector in the case of faults.

Assumption 6.1. *It is assumed that the condition* $\|g_2(\Lambda - I)g_2^{-1}\|_\infty < 1$ *holds.*

Remark 6.1. *[150] There exists a condition* $\|g_2(\Lambda - I)g_2^{-1}\|_\infty < 1$, *such that the control signals* $g_2 u$ *dominate the fault vector function* $g_2(\Lambda - I)u$.

By accounting for the actuator faults and model uncertainties, the control-oriented model in vector format is established as:

$$
\begin{cases}
\dot{x}_1 = g_1 x_2 + \Delta_1 \\
\dot{x}_2 = f_2 + g_2 u + \Delta_2
\end{cases}, \tag{6.16}
$$

where Δ_1 arises from f_1, and $\Delta_2 = g_2(\Lambda - I)u + g_2\rho$ is the lumped uncertainty induced by actuator faults.

By defining $y_1 = x_1 - x_{1,d}$ and $y_2 = g_1 x_2 - \dot{x}_{1,d}$, the following equations can be achieved:

$$
\begin{cases}
\dot{y}_1 = y_2 + \Delta_1 \\
\dot{y}_2 = \dot{g}_1 x_2 + g_1 f_2 - \ddot{x}_{1,d} + g_1 g_2 u + g_1 \Delta_2
\end{cases}, \tag{6.17}
$$

where $x_{1,d}$ represents the desired states. Letting $a(\cdot) = \dot{g}_1 x_2 + g_2 f_2 - \ddot{x}_{1,d}$, $b(\cdot) = g_1 g_2$, and $\Delta_3 = g_1 \Delta_2$, Eq. (6.17) can be simplified as:

$$
\begin{cases}
\dot{y}_1 = y_2 + \Delta_1 \\
\dot{y}_2 = a + bu + \Delta_3
\end{cases}. \tag{6.18}
$$

Remark 6.2. *As can be seen from Fig. 6.1 and also Eq. (6.8), the input vector of the outer-loop HGV model consists of the roll rate (p), pitch rate (q), and yaw rate (r), while the state vector with respect to the outer-loop is composed by the bank angle (μ), AOA (α), and sideslip angle (β), respectively. Focusing on the inner-loop HGV model of Eq. (6.12), the deflections of the aileron (δ_a), elevator (δ_e), and rudder (δ_r) are regarded as the inputs, while the roll rate, pitch rate, and yaw rate constitute the state vector. It should be mentioned that f_2 in Eq. (6.12) is closely related to the states of Eq. (6.8).*

Remark 6.3. *Unpredictable aerodynamics due to hypersonic speed and airframe/ structural dynamics interactions constitute the uncertainty source.*

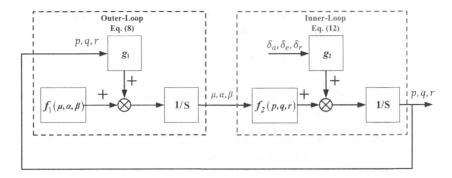

FIGURE 6.1: Block diagram of the HGV model.

As can be seen from Eq. (6.7), the aerodynamic coefficients C_L and C_Y with uncertainty are contained in f_1. Hence, $f_1 = \Delta_1$ in Eq. (6.16) is regarded as the model uncertainty. Moreover, the term in Eq. (6.18), $\Delta_3 = g_1 g_2 (\Lambda - I) u + g_1 g_2 \rho$, includes the information of actuator faults, without a priori knowledge. In the following, fixed-time observers are developed to estimate Δ_1 and Δ_3, respectively.

Remark 6.4. Note that $\Delta_1 = f_1 = [f_\mu, f_\alpha, f_\beta]^T$. In Eq. (6.6), f_μ, f_α, and f_β are regarded as the impact terms of trajectory on the HGV attitude. The value of V is usually very large over HGV flight envelopes. Furthermore, $\beta \approx 0$ and $\gamma \neq \pm\pi/2$. Therefore, the assumption that Δ_1 is bounded is reasonable. Focusing on $\Delta_3 = g_1 g_2 (\Lambda - I) u + g_1 g_2 \rho$, Δ_3 is related to system states and control inputs, which are bounded. With respect to HGV fight envelopes, $\beta \approx 0$ and each element of g_2 is composed by bounded control moment coefficients. In consequence, Δ_3 is bounded in flight.

Remark 6.5. As reported in [151], hydraulic driven actuators are configured in hypersonic vehicles to operate all control surfaces. Flush air data system (FADS) that is often mounted in the upper and lower lifting surfaces has been successfully applied to hypersonic vehicles [152]. FADS, which is dependent on the pressure sensor array measurement of aircraft surface pressure distribution, obtains dynamic pressure, bank angle, AOA, and sideslip angle indirectly through a specific algorithm. In addition, an inertial navigation system (INS) can measure the position, orientation, and velocity of a hypersonic vehicle. Hence, with respect to the studied HGV, the bank angle, AOA, and sideslip angle can be measured by an FADS, while the measurements of the angular rates of roll, pitch, and yaw can be provided by an INS.

6.2.3 Problem Statement

The objective is to design a safety control scheme such that:

1. The terms including actuator faults and system uncertainties can be estimated within a fixed amount of time, thus:

$$\lim_{t \to t_o} \|\hat{\Delta}_1 - \Delta_1\| = 0, \lim_{t \to t_o} \|\hat{\Delta}_3 - \Delta_3\| = 0, \tag{6.19}$$

where t_o is the fixed convergence time, $\hat{\Delta}_1$ and $\hat{\Delta}_3$ are the estimates of Δ_1 and Δ_3, respectively.

2. The detrimental impact of HGV actuator faults can be counteracted within a finite amount of time, thus:

$$\lim_{t \to t_c} |\mu - \mu_d| = 0, \lim_{t \to t_c} |\alpha - \alpha_d| = 0, \lim_{t \to t_c} |\beta - \beta_d| = 0, \tag{6.20}$$

where t_c denotes the finite convergence time, μ_d, α_d, and β_d correspond to the reference signals of the bank angle, AOA, and sideslip angle, respectively.

3. The composite-loop design is achieved under multivariable situation, by which separating the HGV dynamics into inner and outer loops is no longer needed.

6.3 Fixed-Time Observer

6.3.1 An Overview of the Developed Observer and Accommodation Architecture

As illustrated in Fig. 6.2, the developed safety control scheme consists of a fixed-time observer and a finite-time safety control unit. Once the HGV encounters actuator faults, the observer can be continuously run in an effort to obtaining the information within fixed settling time. Then, the safety control responds to the observer results, ensuring the states of the handicapped HGV approach to the desired ones within finite time. Therefore, two problems are addressed in the following. The first is the observer design by means of the fixed-time multivariable sliding mode technique. The second is the synthesis of the safety control against actuator faults and model uncertainties, using the finite-time multivariable integral terminal SMC (TSMC) method.

6.3.2 Fixed-Time Observer

Theorem 6.1. *Consider the faulty system described by Eq. (6.18), and assume that the terms Δ_1 and Δ_3 satisfy $\|\dot{\Delta}_1\| \leq L_1$ and $\|\dot{\Delta}_3\| \leq L_2$, where L_1 and L_2 are known constants. Define z_1, z_2, z_3, and z_4 as the states of the designed fixed-time observers. If the observers are designed by Eqs. (6.21)-(6.22)*

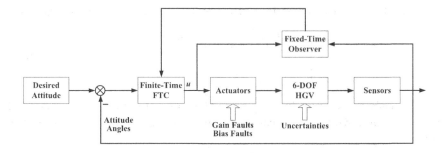

FIGURE 6.2: Conceptual HGV safety control block diagram.

under the condition (6.23), therefore the terms Δ_1 and Δ_3 can be estimated within fixed time through z_2 and z_4, respectively.

$$
\begin{cases}
\dot{z}_1 = -l_1 \dfrac{e_1}{\|e_1\|^{1/2}} - l_2 e_1 \|e_1\|^{p-1} + z_2 + y_2 \\[2mm]
\dot{z}_2 = -l_3 \dfrac{e_1}{\|e_1\|}
\end{cases}, \tag{6.21}
$$

$$
\begin{cases}
\dot{z}_3 = -l_4 \dfrac{e_2}{\|e_2\|^{1/2}} - l_5 e_2 \|e_2\|^{p-1} + z_4 + a + bu \\[2mm]
\dot{z}_4 = -l_6 \dfrac{e_2}{\|e_2\|}
\end{cases}, \tag{6.22}
$$

where $e_1 = z_1 - y_1$, $e_2 = z_3 - y_2$, $p > 1$, and

$$
\begin{cases}
l_1 > \sqrt{2l_3}, l_2 > 0, l_3 > 4L_1 \\[2mm]
l_4 > \sqrt{2l_6}, l_5 > 0, l_6 > 4L_2
\end{cases}. \tag{6.23}
$$

Proof. By taking the derivative of e_1, one can obtain that:

$$
\begin{aligned}
\dot{e}_1 &= \dot{z}_1 - \dot{x}_1 \\
&= -l_1 \frac{e_1}{\|e_1\|^{1/2}} - l_2 e_1 \|e_1\|^{p-1} + z_2 - \Delta_1.
\end{aligned} \tag{6.24}
$$

Letting $e_1^* = z_2 - \Delta_1$, Eq. (6.24) can be rewritten as:

$$
\dot{e}_1 = -l_1 \frac{e_1}{\|e_1\|^{1/2}} - l_2 e_1 \|e_1\|^{p-1} + e_1^*. \tag{6.25}
$$

Taking the derivative of e_1^* gives:

$$
\dot{e}_1^* = \dot{z}_2 - \dot{\Delta}_1 = -l_3 \frac{e_1}{\|e_1\|} - \dot{\Delta}_1. \tag{6.26}
$$

As a result, the error dynamics of the observer for Δ_1 can be represented as:

$$\begin{cases} \dot{e}_1 = -l_1 \dfrac{e_1}{\|e_1\|^{1/2}} - l_2 e_1 \|e_1\|^{p-1} + e_1^* \\ \dot{e}_1^* = -l_3 \dfrac{e_1}{\|e_1\|} - \dot{\Delta}_1 \end{cases}. \tag{6.27}$$

On the basis of the result in [153], when the observer gains l_1, l_2, and l_3 satisfy the condition (23), e_1 and e_1^* can uniformly converge to the origin within fixed time:

$$t_o \leq \left(\frac{1}{l_2(p-1)\varepsilon^{p-1}} + \frac{2(\sqrt{2}\varepsilon)^{1/2}}{l_1} \right)$$
$$\times \left(1 + \frac{l_3 + L}{(l_3 - L)(1 - \sqrt{2l_3}/l_1)} \right), \tag{6.28}$$

where $\varepsilon > 0$. The minimum value of $t_o(\varepsilon)$ is obtained as long as $\varepsilon = (2^{1/4}l_1/l_2)^{\frac{1}{p+1/2}}$. Recalling the definition $e_1^* = z_2 - \Delta_1$, it is proven that z_2 can approach to Δ_1 within fixed time.

Moreover, the proof procedure of the fixed-time observation of Δ_3 is akin to that of Δ_1. For the sake of space, the details are omitted herein. $\qquad\square$

Remark 6.6. *In most of the observer design approaches, the estimation error can vanish asymptotically or within finite time. In contrast, the developed observers are capable of estimating Δ_1 and Δ_3 within fixed time. According to Eqs. (6.21)-(6.22), there are three observer gains for each observer. The purpose of selecting l_i ($i = 1, 2, 3$) is to ensure that the estimation errors ($\dot{e}_1$ and $\dot{e}_1^*$) can converge to the origin within fixed time. To be more specific, the obtained condition $l_3 > 4L_1$ drives $\dot{e}_1^*$ to zero within fixed time. Subsequently, the conditions, $l_1 > \sqrt{2l_3}$ and $l_2 > 0$, enable $\dot{e}_1$ to reach to the origin within fixed time. Hence, this property is important especially for safety-critical aerospace engineering systems.*

6.4   Finite-Time Accommodation Design

With respect to Eq. (6.18), let $\tilde{y}_1 = y_1$ and $\tilde{y}_2 = y_2 + \hat{\Delta}_1$, where $\hat{\Delta}_1$ denotes the estimated value of Δ_1 that can be achieved by the proposed fixed-time observer. Define an integral terminal sliding manifold as:

$$S = \tilde{y}_2 + \int_0^\tau k_1 \|\tilde{y}_1\|^{r_1} \frac{\tilde{y}_1}{\|\tilde{y}_1\|} + k_2 \|\tilde{y}_2\|^{r_2} \frac{\tilde{y}_2}{\|\tilde{y}_2\|} d\tau, \tag{6.29}$$

where $k_1, k_2 > 0$, $r_2 \in (0, 1)$, and $r_1 = 2r_2/(2 - r_2)$. The aim is to drive y_1 to the origin along $S = 0$ in finite time, despite that actuator faults and model uncertainties are present in the HGV.

Theorem 6.2. *The HGV safety control law is formulated as:*

$$u = -b^{-1}(a + k_1 \|\tilde{y}_1\|^{r_1} \frac{\tilde{y}_1}{\|\tilde{y}_1\|} + k_2 \|\tilde{y}_2\|^{r_2} \frac{\tilde{y}_2}{\|\tilde{y}_2\|}$$
$$+ z_4 + \dot{z}_2 + \eta_1 S + \eta_2 \|S\|^{r_3} \frac{S}{\|S\|}), \tag{6.30}$$

where $r_3 \in (0,1)$. Thus, the proposed safety control law ensures that y_1 is steered to the origin along $S = 0$ in finite time, when actuator faults and model uncertainties are present.

Proof. Given the FTC law Eq. (6.30), differentiating the sliding surface Eq. (6.29) along the faulty system Eq. (6.18) can render:

$$\dot{S} = \dot{\tilde{y}}_2 + k_1 \|\tilde{y}_1\|^{r_1} \frac{\tilde{y}_1}{\|\tilde{y}_1\|} + k_2 \|\tilde{y}_2\|^{r_2} \frac{\tilde{y}_2}{\|\tilde{y}_2\|}$$
$$= \dot{y}_2 + \dot{z}_2 + k_1 \|\tilde{y}_1\|^{r_1} \frac{\tilde{y}_1}{\|\tilde{y}_1\|} + k_2 \|\tilde{y}_2\|^{r_2} \frac{\tilde{y}_2}{\|\tilde{y}_2\|}$$
$$= a + bu + z_4 + \dot{z}_2 + k_1 \|\tilde{y}_1\|^{r_1} \frac{\tilde{y}_1}{\|\tilde{y}_1\|} + k_2 \|\tilde{y}_2\|^{r_2} \frac{\tilde{y}_2}{\|\tilde{y}_2\|} \tag{6.31}$$
$$= (z_4 - \hat{z}_4) - \eta_1 S - \eta_2 \|S\|^{r_3} \frac{S}{\|S\|}.$$

By letting $e_{z_4} = \hat{z}_4 - z_4$ and $e_{z_2} = \hat{z}_2 - z_2$, Eq. (6.31) can be simplified as:

$$\dot{S} = -e_{z_4} - \eta_1 S - \eta_2 \|S\|^{r_3} \frac{S}{\|S\|}. \tag{6.32}$$

Define a finite-time bounded function [144]:

$$V_1(S, \tilde{y}_1, \tilde{y}_2) = \frac{1}{2}(S^T S + \tilde{y}_1^T \tilde{y}_1 + \tilde{y}_2^T \tilde{y}_2). \tag{6.33}$$

Note that the parameter r_i ($i = 1, 2$) satisfies the condition $0 < r_i < 1$, which implies that $\|\tilde{y}_i\|^{r_i} < 1 + \|\tilde{y}_i\|$. By taking derivative of V_1 along dynamics Eq. (6.18), one obtains:

$$\dot{V}_1 = S^T \dot{S} + \tilde{y}_1^T \dot{\tilde{y}}_1 + \tilde{y}_2^T \dot{\tilde{y}}_2$$
$$= S^T \dot{S} + \tilde{y}_1^T (\tilde{y}_2 - e_{z_2})$$
$$+ \tilde{y}_2^T (\dot{S} - k_1 \|\tilde{y}_1\|^{r_1} \frac{\tilde{y}_1}{\|\tilde{y}_1\|} - k_2 \|\tilde{y}_2\|^{r_2} \frac{\tilde{y}_2}{\|\tilde{y}_2\|})$$
$$\leq \|S^T e_{z_4}\| + \|\tilde{y}_1\| (\|\tilde{y}_2\| + \|e_{z_2}\|)$$
$$+ \|\tilde{y}_2\| (\|e_{z_4}\| + \eta_1 \|S\| + \eta_2(1 + \|S\|))$$
$$+ \|\tilde{y}_2\| (k_1(1 + \|\tilde{y}_1\|) + k_2(1 + \|\tilde{y}_2\|))$$

$$\leq \frac{\left\|S^T\right\|^2 + \left\|e_{z_4}\right\|^2}{2} + \frac{\left\|\tilde{y}_1\right\|^2 + \left\|\tilde{y}_2\right\|^2}{2} + \frac{\left\|\tilde{y}_1\right\|^2 + \left\|e_{z_2}\right\|^2}{2}$$

$$+ \frac{\left\|\tilde{y}_2\right\|^2 + \left\|e_{z_4}\right\|^2}{2} + \frac{\eta_1 + \eta_2}{2}\left(\left\|\tilde{y}_2\right\|^2 + \left\|S\right\|^2\right) \qquad (6.34)$$

$$+ \frac{\eta_2^2 + \left\|\tilde{y}_2\right\|^2}{2} + \frac{(k_1 + k_2)^2 + \left\|\tilde{y}_2\right\|^2}{2}$$

$$+ \frac{k_1(\left\|\tilde{y}_1\right\|^2 + \left\|\tilde{y}_2\right\|^2)}{2} + k_2 \left\|\tilde{y}_2\right\|^2$$

$$= \left(\frac{1}{2} + \frac{\eta_1 + \eta_2}{2}\right)\left\|S^T\right\|^2 + \left(1 + \frac{k_1}{2}\right)\left\|\tilde{y}_1\right\|^2$$

$$+ \left(2 + \frac{k_1}{2} + \frac{\eta_1 + \eta_2}{2} + k_2\right)\left\|\tilde{y}_2\right\|^2$$

$$+ \left(\left\|e_{z_4}\right\|^2 + \frac{1}{2}\left\|e_{z_2}\right\|^2 + \frac{\eta_2^2}{2} + \frac{(k_1 + k_2)^2}{2}\right)$$

$$\leq K_{V_1} V_1 + L_{V_1},$$

where $K_{V_1} = 4 + k_1 + \eta_1 + \eta_2 + 2k_2$, and $L_{V_1} = \max\left(\left\|e_{z_4}\right\|^2 + \frac{1}{2}\left\|e_{z_2}\right\|^2 + \frac{\eta_2^2}{2} + \frac{(k_1+k_2)^2}{2}\right)$, respectively. *Theorem 6.1* guarantees that the estimation errors e_{z_2} and e_{z_4} converge to zero in fixed time, which implies that e_{z_2} and e_{z_4} are bounded. In addition, L_{V_1} is bounded. Therefore, it can be concluded that V_1 and the state $\tilde{y}_i$ will not escape to infinity before the convergence of the observer error dynamics. Since e_{z_2} and e_{z_4} approach to zero in fixed time, Eq. (6.32) in turn becomes:

$$\dot{S} = -\eta_1 S - \eta_2 \left\|S\right\|^{r_3} \frac{S}{\left\|S\right\|}, \qquad (6.35)$$

which is finite-time stable. As long as the sliding surface is reached, the equivalent dynamics can be obtained using $\dot{S} = 0$:

$$\dot{\tilde{y}}_2 + k_1 \left\|\tilde{y}_1\right\|^{r_1} \frac{\tilde{y}_1}{\left\|\tilde{y}_1\right\|} + k_2 \left\|\tilde{y}_2\right\|^{r_2} \frac{\tilde{y}_2}{\left\|\tilde{y}_2\right\|} = 0. \qquad (6.36)$$

In consequence, it can be proved that the dynamics of Eq. (6.36) is finite-time stabilized, which can be represented as:

$$\begin{cases} \dot{\tilde{y}}_1 = \tilde{y}_2 \\ \dot{\tilde{y}}_2 = -k_1 \left\|\tilde{y}_1\right\|^{r_1} \frac{\tilde{y}_1}{\left\|\tilde{y}_1\right\|} - k_2 \left\|\tilde{y}_2\right\|^{r_2} \frac{\tilde{y}_2}{\left\|\tilde{y}_2\right\|} \end{cases}. \qquad (6.37)$$

Select another Lyapunov function as:

$$V_2 = k_1 \frac{\left\|\tilde{y}_1\right\|^{r_1+1}}{r_1 + 1} + \frac{\left\|\tilde{y}_2\right\|^2}{2}. \qquad (6.38)$$

The time derivative of V_2 along the proceeding dynamics Eq. (6.37) can be written as:

$$
\begin{aligned}
\dot{V}_2 &= k_1 \left\| \tilde{y}_1 \right\|^{r_1-1} \dot{\tilde{y}}_1^T \tilde{y}_1 - \tilde{y}_2^T (k_1 \frac{\tilde{y}_1}{\left\| \tilde{y}_1 \right\|^{1-r_1}} + k_2 \frac{\tilde{y}_2}{\left\| \tilde{y}_2 \right\|^{1-r_2}}) \\
&= k_1 \left\| \tilde{y}_1 \right\|^{r_1-1} \dot{\tilde{y}}_1^T \tilde{y}_1 - k_1 \left\| \tilde{y}_1 \right\|^{r_1-1} \tilde{y}_2^T \tilde{y}_1 \\
&= -k_2 \left\| \tilde{y}_2 \right\|^{r_2-1} \tilde{y}_2^T \tilde{y}_2 \\
&= -k_2 \left\| \tilde{y}_2 \right\|^{r_2-1} \tilde{y}_2^T \tilde{y}_2 \\
&= -k_2 \left\| \tilde{y}_2 \right\|^{r_2+1} .
\end{aligned}
\tag{6.39}
$$

By applying LaSalles invariance principle, the set $\{(\tilde{y}_1, \tilde{y}_2) : \dot{V}(\tilde{y}_1, \tilde{y}_2) = 0\}$ consists of $\tilde{y}_2 = 0$, while the only invariant set inside $\tilde{y}_2 = 0$ is the origin $\tilde{y}_1 = \tilde{y}_2 = 0$. As a result, the asymptotic convergence of $\tilde{y}_1$ and $\tilde{y}_2$ is ensured. Consider the vector field Eq. (6.37) and the dilation [132]:

$$
\begin{aligned}
&(\tilde{y}_{1,1}, \tilde{y}_{1,2}, \tilde{y}_{1,3}, \tilde{y}_{2,1}, \tilde{y}_{2,2}, \tilde{y}_{2,3}) \mapsto \\
&(\kappa \tilde{y}_{1,1}, \kappa \tilde{y}_{1,2}, \kappa \tilde{y}_{1,3}, \kappa^{\frac{1}{2-r_2}} \tilde{y}_{2,1}, \kappa^{\frac{1}{2-r_2}} \tilde{y}_{2,2}, \kappa^{\frac{1}{2-r_2}} \tilde{y}_{2,3}),
\end{aligned}
\tag{6.40}
$$

where $\kappa > 0$. Hence, the vector field Eq. (6.37) is homogeneous of the degree of $\frac{r_2-1}{2-r_2} < 0$. Based on [132], the closed-loop system Eq. (6.36) is globally finite-time stable. In this case, $\tilde{y}_1$ and $\tilde{y}_2$ approach to zero in finite time. □

Remark 6.7. *The reaching phase time and the sliding phase time are finite on the basis of Eq. (6.35) and Eq. (6.37), respectively. Thus, the finite-time stability is successfully incorporated in the safety control against HGV actuator malfunctions and model uncertainties. On the other hand, the control cost from the actuators demanded by the fixed-time control is much larger than that under the finite-time control [154]. With respect to post-fault HGVs, inappropriate control costs may induce the actuator amplitude saturation and even second damage of healthy actuators in the course of actuator fault accommodation. From this fundamental aspect, the* finite-time *control concept is chosen instead of the* fixed-time *control at the safety control design stage.*

Remark 6.8. *Considering that the fault recovery time of an HGV is limited, both the reaching phase time and the sliding phase time are finite in terms of Eq. (6.36) and Eq. (6.38), respectively. This feature is integrated into the safety control design, allowing the states of the faulty HGV to return to the equilibrium within finite time.*

Remark 6.9. *Multivariable SMC design for a hypersonic vehicle is focused in [136]. However, the design is transformed into a decoupled one. Instead, the developed approach of this study has twofold benefits: 1) the problem associated with the decoupled design is avoided; and 2) the safety control is designed, with the assurance that the globally finite-time stability is achieved in spite of actuator faults. These improvements have the potential of enhancing the safety of operational HGVs.*

Remark 6.10. $\dot{z}_2$ *can be obtained by two methods. One option is to obtain* $\dot{z}_2$ *(observer state) directly from the fixed-time observer. However,* $\dot{z}_2$ *is not a continuous signal. Instead,* $\dot{z}_2$ *can be estimated on-line by the robust exact fixed-time differentiator [155]. The differentiator can be implemented if the higher order derivatives of the input are bounded and the fixed-time escape does not exist. The differentiator transient can be driven adequately short by properly tuning the differentiator parameters.*

6.5 Numerical Simulations

6.5.1 HGV Flight Conditions

The initial flight conditions of the selected HGV are: $V(0) = 3000\,\text{m/s}$, $H(0) = 3000\,\text{m}$, $\mu(0) = 0\,^\circ$, $\alpha(0) = 2\,^\circ$, $\beta(0) = 0\,^\circ$, and $p(0) = q(0) = r(0) = 0$. The geometric parameters are: $m = 641.7\,\text{kg}$, $\bar{b} = \bar{c} = 0.8\,\text{m}$, $S_r = 0.5024\,\text{m}^2$, $I_{xx} = 65.12\,\text{kg} \cdot \text{m}^2$, $I_{yy} = 247.26\,\text{kg} \cdot \text{m}^2$, and $I_{zz} = 247.26\,\text{kg} \cdot \text{m}^2$.

6.5.2 Simulation Scenarios

The control without fixed-time observer and the proposed safety control are both examined. The actuator faults, model uncertainties, and measurement noises are considered in the simulation.

1. Actuator time-varying faults:

$$\lambda_1 = \begin{cases} 1 & 0 \leq t < 4 \\ 0.7 - \dfrac{t-4}{5} \times 0.3 & 4 \leq t < 9 \\ 0.4 & t \geq 9 \end{cases},$$

$$\rho_1 = \begin{cases} 0 & 0 \leq t < 4 \\ 5 + \dfrac{t-4}{5} \times 4 & 4 \leq t < 9 \\ 9 & t \geq 9 \end{cases}. \tag{6.41}$$

$$\lambda_2 = \begin{cases} 1 & 0 \leq t < 4 \\ 0.7 - \dfrac{t-4}{5} \times 0.2 & 4 \leq t < 9 \\ 0.5 & t \geq 9 \end{cases},$$

$$\rho_2 = \begin{cases} 0 & 0 \leq t < 4 \\ -9 + \dfrac{t-4}{5} \times 18 & 4 \leq t < 9 \\ 9 & t \geq 9 \end{cases}. \tag{6.42}$$

$$\lambda_3 = \begin{cases} 1 & 0 \le t < 4 \\ 0.7 - \dfrac{t-4}{5} \times 0.2 & 4 \le t < 9 \,, \\ 0.5 & t \ge 9 \end{cases}$$

$$\rho_3 = \begin{cases} 0 & 0 \le t < 4 \\ 9 - \dfrac{t-4}{5} \times 19 & 4 \le t < 9 \,. \\ -10 & t \ge 9 \end{cases}$$

(6.43)

2. According to [156], the maximal degree of the mismatch in the aerodynamic moment coefficients (C_l, C_m, and C_n) is chosen as 20%. The roll, pitch, and yaw moments of inertia (I_{xx}, I_{yy}, and I_{zz}) are perturbed by 20% of the nominal values. Moreover, the maximum dispersion of the selected HGV mass is 20% of the nominal value.

3. The actuator dynamics are chosen as $40/(s + 40)$ in the simulation studies. In addition, the white noise with a mean of 0 and covariance of 0.01 is injected into each measurement channel.

To quantitatively assess the attitude tracking performance, three indices corresponding to μ, α, and β are defined as:

$$\sigma_i = \sqrt{\frac{1}{t_2 - t_1} \int_{t_1}^{t_2} |y_{1,i}|^2 \, d\tau}, \quad i = 1, 2, 3,$$

(6.44)

where $[t_1, t_2]$ covers the time frame of the simulation run and $y_{1,i}$ is the ith element of y_1. The defined metric is the scalar valued L_2 norm, as a measure of average tracking performance [157]. The norm measures the root-mean-squared "average" of the tracking error. A smaller L_2 norm indicates smaller tracking error and thus better tracking performance.

6.5.3 Simulation Results

The tracking performance and tracking errors of the HGV attitude are shown in Fig. 6.3 and Fig. 6.4, respectively. Both the selected safety control schemes can ensure the safety of the faulty HGV. As can be seen from Fig. 6.3(a) and Fig. 6.4(a), the tracking error of bank angle under the developed safety control scheme is significantly smaller than that of the safety control scheme without fixed-time observer. It is illustrated in Fig. 6.3(b) and Fig. 6.4(b) that the AOA tracking performance achieved by the designed safety control is better in comparison to the safety control without fixed-time observer. Fig. 6.3(c) and Fig. 6.4(c) show that the presented scheme outperforms the comparing safety control with respect to sideslip angle. Hence, from Fig. 6.3 and Fig. 6.4, the tracking errors remain remarkably smaller during the entire maneuver when the proposed safety control is used. In terms of the curves, the tracking performance by the safety control without fixed-time observer

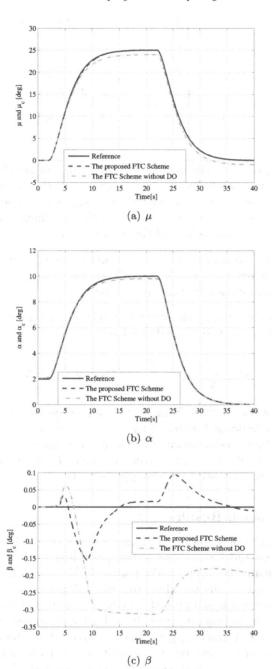

FIGURE 6.3: The responses of μ, α, and β.

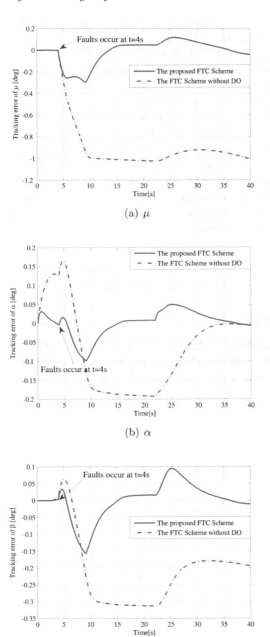

(a) μ

(b) α

(c) β

FIGURE 6.4: The tracking errors of μ, α, and β.

is inferior to that of the proposed safety control. The closed-loop behavior remains excellent in the case of the developed safety control, although the tracking errors exhibit a slightly worse transient behavior. As is visible in Fig. 6.5, the actuators governed by the proposed safety control can satisfactorily handle the time-varying faults.

The improved rate of σ_1 from the safety control without fixed-time observer to the designed safety control is 41.56% (from 0.77° to 0.45°). The measure of σ_2 is enhanced by 47.46% (from 0.59° to 0.31°) when the selected schemes are compared. With respect to σ_3, the studied safety control also attains the superior performance than that of the control without fixed-time observer, with 49.02% improvement of the defined metric (from 0.51° to 0.26°). The performance indices emphasize that the safety control design approach is applicable not only for ensuring the HGV safety, but also for preserving the sound tracking performance.

6.6   Conclusions

A new development of integrating fixed-time observers and finite-time control into a safety control scheme is presented for a HGV, which can handle actuator faults and model variations. The benefits of the algorithms include:

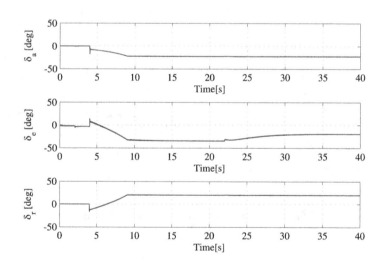

FIGURE 6.5: Control signals δ_a, δ_e, and δ_r.

1) the estimation errors can be driven to zero within fixed time; 2) the safety control law is constructed using the composite-loop design concept, by which the finite-time stability of the faulty closed-loop system can be ensured; and 3) multivariable situation is incorporated at the design stages of both observer and safety control for avoiding the decoupled issues induced by formulating a multi-input and multi-output system as m single-input systems. These improvements offer the potential to enhance the safety of hypersonic vehicles. The case studies based on a full nonlinear model of the HGV dynamics demonstrate that the proposed safety control scheme can effectively deal with scenarios involving actuator faults and model variations.

6.7 Notes

The main contribution of this chapter is to design a fixed-time fault-tolerant control scheme for a hypersonic gliding vehicle to counteract actuator faults and model uncertainties. In contrast to [136, 144, 145], the convergence time of the designed disturbance observer is independent of initial conditions, while elegant solutions can be achieved by vector form design, which is more suitable for the HGV safety requirement. In addition to observer design aspect, this study explicitly considers "mismatched uncertainty" which is used to handle the condition of actuator faults and model uncertainties, while this type of uncertainty is not prescribed in the recent work [145].

Chapter 7

Fault Accommodation with Consideration of Control Authority and Gyro Availability

7.1 Introduction

Flight actuators, such as elevators, ailerons, and rudders, play a central role in any flight control system (FCS). They are capable of either altering aircraft maneuvers or attenuating external disturbances, as long as adequate control authority exists. In real aerospace applications, exhaustion of physical control authority is known as magnitude and/or rate limiting of the actuators. To be more specific, an actuator has displacement authority mainly due to an actuation system's travel limits. Actuator amplitude saturation is dangerous, as it does not only affect aircraft performance but also flight safety. Additionally, actuation rate bounds on a flight actuator are inevitable owing to the maximum flow capability of the main control valve and the flight-induced aerodynamic load. It is a well-established fact that the actuation rate limiting is chiefly attributable to pilot-induced oscillations (PIOs), as shown from the preceding flight tests of the C-17 and Boeing 777 transport aircraft [158, 159]. It is also reported that actuator rate saturation was implicated in deteriorating aircraft's safety [160, 161, 162]. As a consequence, the study on saturation of nonlinear behaviors has been a major focus in FCS design over many years. Under normal flight conditions, a variety of control methods attempt to address the saturation of actuator amplitude [163, 164, 165, 166, 167] and rate [168, 169, 170], respectively. A discussion of developing a flight control law is presented in Ref. [171], addressing the challenging problem which involves both the magnitude and rate bounds of flight actuators. Based on sector conditions, a new linear matrix inequality characterization is proposed to achieve anti-windup compensators against both magnitude and rate limitations.

As actuator faults are detected and identified (e.g., by an observer-based method [172]), an increasing need to use magnitudes (and associated actuation rates) of healthy actuators is required. The healthy actuators may therefore take a higher risk of control authority exhaustion than usual [133]. The fault-free actuators may be over-stretched to counteract actuator malfunctions if a fault accommodation scheme is improperly designed. This situation can result in secondary damage or aircraft break-down. Hence, research on aircraft fault accommodation taking into consideration of the actuator physical capability has attracted growing attention. The literature concerning this topic can be dominated by two concepts: (1) the magnitude limits of redundant actuators are not violated by properly managing command/reference signals [11, 49, 50, 51]. The performance requirements from both transient and steady-state aspects are gracefully degraded, by which the remaining actuators are not aggressively driven. This sort of method can be thought as an indirect strategy; (2) the control laws are reconfigured with integration of actuator displacement bounds. In Ref. [52], the proposed control is able to size the stability domain according to the actuator magnitude saturation. The invariant ellipsoidal set technique is incorporated into the flight control design for unmanned aerial vehicles (UAVs) [53, 54]. The "positive μ-modification" technique and the adjustable reference model are deployed for preventing remaining actuators from amplitude saturation [55, 56]. As can be summarized from the included findings [52, 53, 54, 55, 56], the problem of actuator amplitude limiting is addressed in a direct manner.

Sensors onboard are of great importance due to the fact that appropriately measured signals must be used for FCSs and flight health management systems (FHMSs). Sensors are identified as the weak link in aerospace engineering systems based on being more vulnerable to damage or being more sensitive in construction than other components [57]. Irrecoverable flight status may be caused by conveying measurements from abnormal sensors into the control loop. Catastrophic consequences can also be triggered if incorrectly sensed information is delivered to an FHMS. To meet the stringent demands of aircraft reliability, physical redundancy in the form of identical sensors or multiple types of sensors is often configured [58]. Nonetheless, redundant sensors are not always feasible due to the constraints of weight, cost, space, power, and complexity. Thereby, analytical redundancy constituted by the knowledge of the aircraft becomes a viable supplement [59].

Angular rate sensors (also named gyroscopes or gyros) are the primary measurement units in FCSs. The failure rate of gyros is ranked the highest among the key components in an inertial measurement unit (IMU) [58]. With respect to gyros, the research outcomes underscore the analytical redundancy's potential of offering reliable angular rates [14, 173, 174, 175, 176]. The reconstructed angular rates are used for an advanced tailless aircraft [173]. Considering additive faults [174] and incipient faults [14] in aircraft gyros, the pitch rate signal is reconstructed using a sliding mode observer (SMO). The tracking problem for vertical take-off and landing (VTOL) aircraft with sensorless angular velocities is studied in Ref. [175], where an observer-based approach is exploited to estimate the angular rates. More recently, the fully connected cascade neural network (NN) architecture is adopted to detect and recover the aircraft gyro failures [176].

Despite previous attempts having achieved various degrees of success in solving fault-tolerance issues, there still exist some challenges: (1) from the literature, the developed systems can accommodate aircraft malfunctions and avoid actuator amplitude saturation. However, care must be taken to ensure actuation rates remain within the allowable bounds for more realistic applications; (2) one cannot guarantee that the closed-loop system is stabilized when the measured gyro signals are substituted by the estimated ones in the feedback loop. The stability of the closed-loop system with reconstructed angular rates needs to be proved; and (3) the vast majority of the developed schemes focus on either actuator fault tolerance or gyro fault reconstruction. Nevertheless, the simultaneous treatment can be more applicable.

Motivated by the aforementioned factors, actuator fault accommodation is investigated with particular attention to the physical control authority, regardless of whether or not the measurable angular rate is present. The main contributions of this chapter lie in two aspects: (1) in the face of actuator failures, a fault accommodation system is proposed based on an adaptive sliding mode control (SMC) technique, where both the displacement and rate limits of healthy actuators are considered in an explicit manner; (2) in addition to actuator failures, an extension of the first presented algorithm is developed to handle a severe situation, where the sensed angular velocity is inaccessible. A SMO-based accommodation strategy is designed, capable of estimating the angular rate and respecting the control authority of fault-free actuators. The SMO is exploited to guarantee that the aircraft angular rate can be reconstructed and subsequently transmitted into the accommodation procedure. Moreover, an integrated design of the SMO and the accommodation is developed such that the faulty aircraft can be stabilized with the recovered angular rate. Thus, an option is offered under such a severe situation.

The rest of this chapter is arranged as follows. The aircraft nonlinear dynamics, actuator amplitude and rate bounds, actuator failure modes, gyro analysis, and problem statement are described in Chapter 7.2. A fault accommodation scheme is proposed against actuator failures in Chapter 7.3, where both the displacement and rate limits of healthy actuators are incorporated. Chapter 7.4 presents a SMO-based accommodation strategy to handle both actuator and sensor gyro failures, where the physical bounds of healthy actuators are also accounted for. The simulation results and analysis are provided in Chapter 7.5 to validate the effectiveness of the proposed approaches. Conclusion remarks and future works are given in Chapter 7.6.

7.2   Aircraft Model and Problem Statement

7.2.1   Longitudinal Aircraft Model Description

Transport category aircraft usually have less excess control authority and higher risk of actuator saturation, rather than highly maneuverable fighters. The Boeing 747, a wide-body commercial jet airliner and cargo aircraft, possesses multiple physical redundancies, allowing the demonstration of accommodation concepts [177]. Thereby, the Boeing 747 benchmark aircraft is selected in this work.

The dynamics of the Boeing 747 aircraft is highly nonlinear. Under the rigid assumption, the longitudinal motion equations can be described as [178, 179]:

$$\dot{V} = \left[-D + T \cos\left(\alpha + \sigma_T\right) - mg \sin\gamma \right]/m, \tag{7.1}$$

$$\dot{\alpha} = \left[-L - L \sin\left(\alpha + \sigma_T\right) + mg \cos\gamma \right] (mV) + q, \tag{7.2}$$

$$\dot{\theta} = q, \tag{7.3}$$

$$\dot{q} = \left[M + T l_{tz} \cos\sigma_T \right]/I_y, \tag{7.4}$$

where V, α, γ, θ, and q denote the true airspeed, angle of attack (AOA), flight path angle (FPA), pitch angle, and pitch rate, respectively. m, g, I_y, T, l_{tz}, and σ_T represent mass, acceleration of gravity, moment of inertia about y-body axis, thrust force, the distance from the engine center line to the fuselage reference line, and the engine inclination angle, respectively.

The drag D, lift L, and pitching moment M can be calculated as:

$$D = \bar{q} S_W C_D, \tag{7.5}$$

$$L = \bar{q} S_W C_L, \tag{7.6}$$

$$M = \bar{q} S_W \bar{c} \left[C_m + \Delta \right], \tag{7.7}$$

where

$$\Delta = -\bar{x}_{cq} \left[C_D \sin\alpha + C_L \cos\alpha \right]/\bar{c} + \bar{c}\dot{\alpha} C_{m\dot{\alpha}}/V, \tag{7.8}$$

and $\bar{q}$ is the dynamic pressure represented by:

$$\bar{q} = \rho V^2/2. \tag{7.9}$$

In Eqs. (7.5)–(7.9), S_w, $\bar{c}$, ρ, and $\bar{x}_{cq}$ specify the wing area, mean aerodynamic chord, air density, and the difference between the actual and the reference x-axis center of gravity. C_D, C_L, and C_m are the drag force coefficient, lift coefficient, and pitching moment coefficient, respectively. $C_{m\dot{\alpha}} = \partial C_m/\partial\dot{\alpha}$ denotes the variation of C_m with respect to $\dot{\alpha}$.

Assumption 7.1. *It is assumed that V keeps constant, which means that $\dot{V} \approx 0$.*

Assumption 7.2. *It is assumed that the elevators mainly affect the pitching moment. The effect on drag and lift can be ignored.*

Remark 7.1. *Assumption 7.1 can be achieved if an individual feedback loop is designed based on the measured speed and on the auto throttle [178]. For transport aircraft, the pitch moment coefficient with respect to the elevator is significantly larger than the lift coefficient and the drag coefficient corresponding to the elevator. Thus, Assumption 7.2 is made to simplify the longitudinal dynamics for design purposes.*

The relationship between the FPA and AOA can be cast as:

$$\gamma = \theta - \alpha, \tag{7.10}$$

yielding $\dot{\gamma} = \dot{\theta} - \dot{\alpha}$.

By applying the preceding assumptions and Eqs. (7.5)–(7.10), the longitudinal dynamics of the Boeing 747 benchmark model can be given by:

$$\dot{\gamma} = [\bar{q}S_W C_L + T\sin(\theta - \gamma + \sigma_T) - mg\cos\gamma]/(mV), \tag{7.11}$$

$$\dot{\theta} = q, \tag{7.12}$$

$$\dot{q} = [\bar{q}S_W \bar{c}(C_m + \Delta) + Tl_{tz}\cos\sigma_T]/I_y. \tag{7.13}$$

The pitching moment coefficient is expressed by:

$$C_m = C_{m0} + (\partial C_m/\partial\delta_{hs})\,\delta_{hs} + (\partial C_m/\partial\delta_{re})\,\delta_{re} + (\partial C_m/\partial\delta_{le})\,\delta_{le}, \tag{7.14}$$

where δ_{hs}, δ_{re}, and δ_{le} denote the deflections of horizontal stabilizer (HS), right elevator (RE), and left elevator (LE), respectively. A trimmable HS is normally configured in the majority of transport category aircraft. The trimmable HS, which does not move in response to control commands, usually contributes to balancing force to maintain the aircraft's equilibrium flight. Instead, the independent elevators serving to change pitch motions are classified as the primary actuator channels. Furthermore, defining $\chi = [\gamma, \theta, q]^T$ and then combining the terms of C_{m0} and Δ yields,

$$C_m^*(x) = C_{m0}(x) - \bar{x}_{cg}[C_D\sin\alpha + C_L\cos\alpha]/\bar{c} + \bar{c}\dot{\alpha}C_{m\dot{\alpha}}/V. \tag{7.15}$$

It should be mentioned that the term $C_{m0}(x)$ is dependent on the defined state, which is detailed in Ref. [180]. Therefore, the longitudinal motion equations can be rewritten as:

$$\begin{bmatrix} \dot{\gamma} \\ \dot{\theta} \\ \dot{q} \end{bmatrix} = \begin{bmatrix} (\bar{q}S_w C_L + T\sin(\theta - \gamma + \sigma_T) - mg\cos\gamma)/(mV) \\ q \\ [\bar{q}S_w\bar{c}(C_m^*(x) + (\partial C_m/\partial\delta_{hs})\,\delta_{hs}) + Tl_{tz}\cos\sigma_T]/I_y \end{bmatrix}$$
$$+ \begin{bmatrix} 0 & 0 \\ 0 & 0 \\ \bar{q}S_w\bar{c}(\partial C_m/\partial\delta_{re})/I_y & \bar{q}S_w\bar{c}(\partial C_m/\partial\delta_{le})/I_y \end{bmatrix} \begin{bmatrix} \delta_{re} \\ \delta_{le} \end{bmatrix}. \tag{7.16}$$

To facilitate the subsequent design procedures, Eq. (7.16) is simplified as:

$$\begin{bmatrix} \dot{x}_1 \\ \dot{x}_2 \end{bmatrix} = \begin{bmatrix} f_1(x) \\ f_2(x) \end{bmatrix} + \begin{bmatrix} 0_{2\times2} \\ g(x) \end{bmatrix} u, \tag{7.17}$$

where $x_1 = [\gamma, \theta]^T, x_2 = q, u = [\delta_{re}, \delta_{le}]^T, g(x) = [\ g_1,\ g_2\ ] = [\bar{q}S_w\bar{c}\,(\partial C_m/\partial\delta_{re})/I_y,\ \bar{q}S_w\bar{c}(\partial C_m/\partial\delta_{le})/I_y]$,
$f_1(x) = \begin{bmatrix} (\bar{q}S_W C_L + T\sin(\theta - \gamma + \sigma_T) - mg\cos\gamma)/(mV) \\ q \end{bmatrix}$ and $f_2(x) = [\bar{q}S_w\bar{c}\ (C_m^*(x)) + (\partial C_m/\partial\delta_{hs})\delta_{hs} + Tl_{tz}cos\sigma_T]/I_y$. In the above equations, the aerodynamic coefficients and the corresponding derivatives can be obtained from wind tunnel and flight tests [180].

7.2.2   Analysis of Flight Actuator Constraints

Flight actuators are imperative components in any FCSs, linking commands to control surface movements. Hydraulic actuation systems capable of producing the immense force are widely configured in civil and military aircraft, such as the Boeing 747 type of aircraft and F-16 fighter. A schematic of a hydraulically driven flight actuator is illustrated in Fig. 7.1. A differential pressure in the control valve is determined in response to the control command $u_{c,i}$. Hydraulic fluid, which is pumped from a reservoir and cleaned by a filter, passes through the control valve. The fluid flow is then pushed into one side of the actuator cylinder and is extracted from another, leading to differential pressure across the piston. As a result, the piston motion can be actuated. Due to the fact that the control surface is mechanically connected to the piston rod via the hinge, the control surface deflection u_i can be driven by the piston displacement.

The selection of hydraulically driven flight actuators solely relies on the performance requirements from both transient and steady-state perspectives. For instance, how fast a control surface can arrive at the expected position; how much force/torque can be provided by rotating a control surface. Once the actuator size is decided, the physical control authority of the chosen actuator can be known accordingly. Two types of constraints normally apply to the control surface. As shown in Fig. 7.1, the maximum stroke of the piston rod is limited due to the fixed chamber size. Consequently, the rotating bounds of the control surface are determined by the rod travel limits and the hinge length. Moreover, the critical factors affecting the movement speed of the piston are

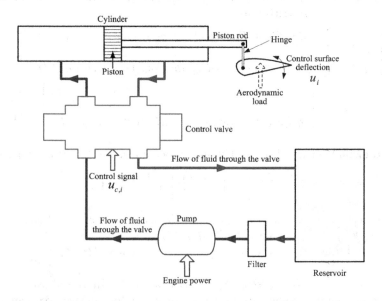

FIGURE 7.1: Schematic diagram of a hydraulically driven flight actuator.

identified to be the piston area, maximum control valve displacement, supply and return pressures, flow rate through the control valve, cross-piston leakage, and flight-induced aerodynamic load [30, 68]. Hence, the deflection rate of the control surface is constrained as well.

The relationship between the control command $u_{c,i}$ and the control surface deflection u_i is approximated by a first-order system [181]:

$$\frac{u_i(s)}{u_{c,i}(s)} = \frac{a_i}{s + a_i}. \tag{7.18}$$

According to Eq. (7.18), one can obtain:

$$u_i = u_{c,i} - a_i^{-1} u_{r,i}, \tag{7.19}$$

where $u_{r,i}$ denotes the deflection rate of the ith control surface. Suppose the amplitude and rate for the ith control surface are symmetrically bounded by $\bar{u}_i$ and $\bar{u}_{r,i}$, respectively. By recalling the representation of the deflection limit and considering the deflection rate constraint, the following equations can be achieved:

$$u_i = \phi_i \cdot \left(u_{c,i} - a_i^{-1} u_{r,i} \right), \tag{7.20}$$

$$u_{r,i} = \text{sgn}(u_{r,i}) \min \{ |u_{r,i}|, \bar{u}_{r,i} \}, \tag{7.21}$$

where ϕ_i s an indicator for the saturation degree is formed by:

$$\phi_i = \begin{cases} 1 & if \ |u_i| \le \bar{u}_i \\ \bar{u}_i \text{sgn} \left(u_{c,i} - a_i^{-1} u_{r,i} \right) / \left(u_{c,i} - a_i^{-1} u_{r,i} \right) & otherwise \end{cases}, \tag{7.22}$$

and *sgn* function is defined as:

$$\text{sgn}(u_{c,i}) = \begin{cases} 1 & if \ u_{r,i} > 0 \\ 0 & if \ u_{r,i} = 0. \\ -1 & if \ u_{r,i} < 0 \end{cases} \tag{7.23}$$

It should be emphasized that the indicator φ_i satisfies $0 < \phi_i \le 1$. When $\phi_i = 1$, the ith actuator works within the amplitude range; while φ_i approaches 0 from 1, the ith actuator suffers from the amplitude saturation with an increasing degree of severity.

Without loss of generality, by applying the physical constraints to multiple flight actuators, Eq. (7.17) can be rewritten as:

$$\dot{x} = \begin{bmatrix} f_1(x) \\ f_2(x) \end{bmatrix} + \begin{bmatrix} 0_{2\times2} \\ g(x) \end{bmatrix} \phi \left(u_c - a^{-1} u_r \right), \tag{7.24}$$

where $\phi = diag\{\phi_1, \phi_2\}$, $u_c = [u_{c,1}, u_{c,2}]^T$, $a = diag\{a_1, a_2\}$, and $u_r = [u_{r,1}, u_{r,2}]^T$, respectively.

7.2.3   Failure Modes and Modeling of Flight Actuators

As long as flight actuators operate normally, the desired deflections can be driven exactly as the FCS has mandated. However, when actuators fail to operate, the control commands cannot be completed as required, and catastrophic consequences may be induced. Based on the fault analysis of a hydraulically driven actuator in Ref. [30], the ith actuator fault model can be presented as:

$$u_i^F = l_i u_i, \tag{7.25}$$

where the factor l_i is declared to indicate the effectiveness of the ith actuator. More particularly, (1) $l_i = 1$ describes that the ith actuator is working under a normal condition; (2) $0 < l_i < 1$ denotes a partial loss of effectiveness in the ith actuator; and (3) $l_i = 0$ represents that a complete failure occurs in the ith actuator. A diagonal matrix L describes the effectiveness in any of the actuators in Eq. (7.24) as:

$$u^F = Lu. \tag{7.26}$$

As a result, the aircraft longitudinal model with actuator faults and limited control authority can be represented as:

$$\dot{x} = \begin{bmatrix} f_1(x) \\ f_2(x) \end{bmatrix} + \begin{bmatrix} 0_{2\times2} \\ g(x) \end{bmatrix} L\phi\left(u_c - a^{-1}u_r\right). \tag{7.27}$$

Remark 7.2. *It is worth mentioning that actuator stuck and runaway are typical failures in hydraulic-type actuators as well. When an actuator is lost due to stuck failure, the faulty actuator is "frozen" at a specific position, which can no longer respond to the applied command. Runaway failure is characterized as when the surface locks at its limited deflection amplitude. It is seen as the worst case of jammed failures. In comparison to the investigated fault in Eq. (7.26), the jammed and runaway failures can impose an external disturbance on the faulty aircraft, deteriorating the safety. Nonetheless, the strategy against jammed and runaway failures is beyond the scope of this chapter.*

7.2.4   Failure Modes and Modeling of Flight Sensor Gyros

According to either the law of conservation of momentum or Sagnac effect, there are two sorts of gyros: mechanical and optical gyros. Conventional spinning gyro and vibrating gyro fall into the mechanical type, while ring laser gyro (RLG), and fibre optic gyro are recognized as the optical type. Conventional spinning gyros are still used for the Boeing 747 transport aircraft [182], albeit with the growing configuration of RLGs in civil aircraft.

As can be observed from [183], the rate gyro consists of (1) the spin motor and rotor; (2) the gimbal; (3) the torsion spring; and (4) the pickoff (output signal generator). The motor converts electrical energy to mechanical energy, which activates the inertia wheel rotating at a constant speed. The motor and the inertia wheel are contained in the gimbal. The springs can provide the rotational resistance to balance the output torque caused by the gyro. The pickoff determines an AC voltage proportional to the rotating speed resolved by the gyro.

Quoting from [58], gyros have the highest failure rates among the crucial components in an IMU. Gyro malfunctions are pertinent to bearing failure induced by the instrument ingesting dirty air, and/or impact damage to the sensitive gyro rotor and gimbal bearings. Inadequate vacuum or pressure system air filtration can accelerate bearing wear. In faulty cases, the sensed angular rates are no longer reliable for FCS and FHMS.

The aircraft output vector can be expressed as:

$$y = h(x), \tag{7.28}$$

where $h(x) : \mathcal{R}^3 \to \mathcal{R}^2$ is a smooth vector field of x.

When sensors suffer faults, Eq. (7.28) can be further described as:

$$y = \Gamma h(x) + \gamma, \tag{7.29}$$

where $\Gamma = diag\{\gamma_1, \gamma_2\}$ and $\gamma = [v_1, v_2]^T$. The typical sensor faults are represented by Eq. (7.29) [57]. (1) When $\gamma_j = 1$ and $v_j = a$ constant offset value, a bias fault occurs in the jth sensor; (2) when $\gamma_j = 1$ and $v_j = a$ time-varying offset factor, a drift fault takes place in the jth sensor; (3) when $\gamma_j = 0$ and $v_j = 0$, the jth sensor suffers from signal loss; and (4) when $\gamma_j = 0$ and $v_j = a$ constant value, the jth sensor encounters a stuck failure.

7.2.5 Problem Statement

As the main contributions of this work, two fault accommodation strategies are developed in response to different fault cases. In regard to the first architecture, the objective is stated as follows:
1) Stabilize the aircraft and retain the tracking performance in the event of actuator failures;
2) The amplitude and rate limits of healthy actuators are not violated when overcoming the deleterious effects of actuator failures.

The second algorithm is designed such that:
1) The aircraft can be stabilized and the tracking performance can be preserved in the case of both actuator failures and gyro failures;
2) The amplitude and rate limits of healthy actuators are respected during the fault accommodation procedure.

7.3   Fault Accommodation with Actuator Constraints

7.3.1   An Overview of the Fault Accommodation Scheme

A general structure of the developed fault accommodation against actuator malfunctions is depicted in Fig. 7.2. The overall system is compounded by an FDD scheme and an adaptive SMC-based accommodation scheme. The real-time information of actuator failures is provided by the FDD unit. Subsequently, the adaptive SMC-based accommodation unit is proposed such that the aircraft outputs can track the reference signals under actuator failures. By virtue of the proposed system, the flight actuator failures can be tolerated within the control authority of fault-free actuators. As a new contribution to the field of safety flight control systems, the design proposed in this study, incorporating the integral SMC and adaptive techniques, can achieve the attitude tracking and handle the actuator bounds on both amplitude and rate after actuator failure occurrences.

7.3.2   Fault Accommodation within Actuator Control Authority

For the aircraft longitudinal model under actuator failures in Eq. (7.27), suppose the reference signals are x_d. According to [184], the sliding surface is selected as:

$$S = G\left(\dot{x} - \dot{x}_d\right) + G \int_0^t \eta\left(\dot{x} - \dot{x}_d\right) dt, \qquad (7.30)$$

where $G \in \mathfrak{R}^{1\times 3}$ stands for the design freedom [178] and $\eta > 0$ is a design parameter for the integral SMC. When $G = [0, 1, 0]$ is chosen, the goal becomes

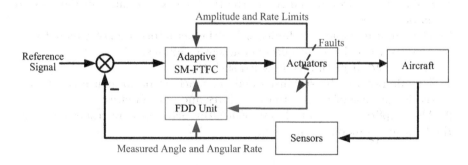

FIGURE 7.2: Illustrative diagram of the developed fault accommodation scheme.

$\theta \to \theta_d$ and the sliding surface turns to be:

$$S = \left(\dot{\theta} - \dot{\theta}_d\right) + \int_0^t \eta \left(\dot{\theta} - \dot{\theta}_d\right) dt. \tag{7.31}$$

Compared to the conventional SMC, an advantage of the above sliding surface with integral action is that the system trajectory always starts from the sliding surface. Accordingly, the reaching phase is eliminated and robustness in the whole state space is maintained. The trajectories of the faulty aircraft (7.27) can be driven on the sliding surface (7.31) and can evolve along it if a proper safety control law u_c is designed. As a consequence, the objective $\theta \to \theta_d$ can be fulfilled for the faulty aircraft.

Note that φ_i (satisfying $0 < \varphi_i \le 1$) is an indicator for describing the degree of saturation severity. The following assumptions are made before presenting the design procedure.

Assumption 7.3. *The fault to be handled in this study is recoverable.*

Assumption 7.4. *There exists one positive constant μ_1 satisfying $0 < \mu_1 \le \min_i \left(g_i \phi_i g_i^{-1}\right)$.*

Assumption 7.5. *There is a positive constant ε such that the condition $\max_i \left(\left|g_i \phi_i a_i^{-1}\right|\right) \le \varepsilon$ holds.*

Assumption 7.6. *It is assumed that $\max_i \left(\bar{u}_{r.i}\right) \le \bar{u}_r$.*

Remark 7.3. *It is known that $g_1 = \bar{q} S_w \bar{c} \left(\partial C_m / \partial \delta_{re}\right) / I_y$, and $g_2 = \bar{q} S_w \bar{c} \left(\partial C_m / \partial \delta_{le}\right) / I_y$ based on the derivation procedure of the simplified longitudinal dynamics. g_1 and g_2 are nonzero and invertible since the variations of C_m with δ_{re} and δ_{le} are nonzero in the typical regions of the flight envelope [181]. The positive parameter a_i in Eq. (7.18) is to represent the characteristics of the ith actuator. Therefore, a_i is also invertible.*

The actuator effectiveness indicators l_1 and l_2 need to be estimated in practice. The imperfect estimation of l_1 and l_2 can deteriorate fault-tolerant performance and even aircraft safety. This factor is thereby considered at the design stage. Based on Ref. [178], the relationship between the true values of the actuator effectiveness factors and the estimated ones can be established as:

$$\begin{cases} l_1 = \hat{l}_1 \left(1 + \delta_1\right) \\ l_2 = \hat{l}_2 \left(1 + \delta_2\right) \end{cases}, \tag{7.32}$$

where $\hat{l}_1$ and $\hat{l}_2$ denote the estimated values of l_1 and l_2, δ_1 and δ_2 specify the inaccurate degrees corresponding to l_1 and l_2, respectively. The scalars δ_1 and δ_2 satisfy:

$$\underline{\delta} \le \delta_1, \delta_2 \le \bar{\delta}, \tag{7.33}$$

where $\underline{\delta}$ and $\bar{\delta}$ are known scalars. Define $\delta_M = \max\left\{\left|\underline{\delta}\right|, \left|\bar{\delta}\right|\right\}$, which is less than 1. Note that $\hat{l}_1$ and $\hat{l}_2$ can be limited in the range [0,1] in a common practice.

Theorem 7.1. *Consider the actuator physical limits imposed by Eqs. (7.20)-(7.21) and actuator failures in Eq. (7.26), the longitudinal dynamics of the Boeing 747 aircraft can be represented by Eq. (7.27) with the sliding surface stated in Eq. (7.31). The trajectory of the faulty closed-loop system can be driven onto the sliding surface S=0 and it can eventually converge to the origin, by applying the following safety control law:*

$$\begin{cases} u_{c,1} = \left[K_1 - g_1^{-1}\hat{\rho}\mathrm{sgn}\,(S)\right] / \left(\hat{l}_1 + \hat{l}_2\right) \\ u_{c,2} = \left[K_2 - g_2^{-1}\hat{\rho}\mathrm{sgn}\,(S)\right] / \left(\hat{l}_1 + \hat{l}_2\right) \end{cases}, \quad (7.34)$$

where

$$\begin{cases} K_1 = \phi_1^{-1}g_1^{-1}\left[\ddot{\theta}_d - \eta\left(\dot{\theta} - \dot{\theta}_d\right) - f_2\,(x)\right] \\ K_2 = \phi_2^{-1}g_2^{-1}\left[\ddot{\theta}_d - \eta\left(\dot{\theta} - \dot{\theta}_d\right) - f_2\,(x)\right] \end{cases}. \quad (7.35)$$

In addition, the estimation of the modulation gain is: where

$$\hat{\rho} = \hat{\rho}_0 + \hat{\rho}_1\,|q| + \hat{\rho}_2\,|f_2^*\,(x)| + \hat{\varepsilon}\bar{u}_r, \quad (7.36)$$

and the update laws are:

$$\dot{\hat{\rho}}_0 = c_0\mu_1\,|S|\,, \quad (7.37)$$

$$\dot{\hat{\rho}}_1 = c_1\mu_1\,|S|\,|q|\,, \quad (7.38)$$

$$\dot{\hat{\rho}}_2 = c_2\mu_1\,|S|\,|f_2^*\,(x)|\,, \quad (7.39)$$

$$\dot{\hat{\varepsilon}} = c_3\mu_1\,|S|\,\bar{u}_r, \quad (7.40)$$

where $\hat{\rho}$, $\hat{\rho}_0$, $\hat{\rho}_1$, $\hat{\rho}_2$, and $\hat{\varepsilon}$ denote the estimated values of the positive constants ρ, ρ_0, ρ_1, ρ_2, and ε and $\rho_2 \geq \delta_M/\mu_1$, $c_j > 0\,(j = 0,\cdots,3)$ are design parameters, and $f_2^*\,(x) = \ddot{\theta}_d - \eta\left(\dot{\theta} - \dot{\theta}_d\right) - f_2\,(x)$.

Proof. Choose a Lyapunov candidate as:

$$\begin{aligned} V = {} & S^2/2 + (\rho_0 - (1+\underline{\delta})\hat{\rho}_0)^2/(2(1+\underline{\delta})c_0) \\ & + (\rho_1 - (1+\underline{\delta})\hat{\rho}_1)^2/(2(1+\underline{\delta})c_1) \\ & + (\rho_2 - (1+\underline{\delta})\hat{\rho}_2)^2/(2(1+\underline{\delta})c_2) \\ & + (2\varepsilon/\mu_1 - (1+\underline{\delta})\hat{\varepsilon})^2/(2(1+\underline{\delta})c_3), \end{aligned} \quad (7.41)$$

when defining $V_1 = S^2/2$, $V_2 = (\rho_0 - (1+\underline{\delta})\hat{\rho}_0)^2/(2(1+\underline{\delta})c_0)$, $V_3 = (\rho_1 - (1+\underline{\delta})\hat{\rho}_1)^2/(2(1+\underline{\delta})c_1)$, $V_4 = (\rho_2 - (1+\underline{\delta})\hat{\rho}_2)^2/(2(1+\underline{\delta})c_2)$, and $V_5 = (2\varepsilon/\mu_1 - (1+\underline{\delta})\hat{\varepsilon})^2/(2(1+\underline{\delta})c_3)$, the selected Lyapunov function can be further divided as:

$$V = V_1 + V_2 + V_3 + V_4 + V_5. \quad (7.42)$$

For the ease of the proof procedure, each term of the right hand side of Eq. (7.42) is individually analyzed.

Step 1) The time derivation of V_1 is:

$$\dot{V}_1 = S\dot{S}. \tag{7.43}$$

According to Eq. (7.31), the time derivative of the sliding surface is:

$$\dot{S} = \ddot{\theta} - \ddot{\theta}_d + \eta\left(\dot{\theta} - \dot{\theta}_d\right). \tag{7.44}$$

Substituting the faulty model (7.27) and the control law (7.34) into Eq. (7.43) results in:

$$\begin{aligned}
\dot{V}_1 =\ & S\left[f_2\left(x\right) - \ddot{\theta}_d + \eta\left(\dot{\theta} - \dot{\theta}_d\right)\right] \\
& + Sg_1\phi_1 l_1 K_1 / \left(\hat{l}_1 + \hat{l}_2\right) + Sg_2\phi_2 l_2 K_2 / \left(\hat{l}_1 + \hat{l}_2\right) \\
& - Sg_1\phi_1 l_1 g_1^{-1}\hat{\rho}\,\mathrm{sgn}\left(S\right)/\left(\hat{l}_1 + \hat{l}_2\right) \\
& - Sg_2\phi_2 l_2 g_2^{-1}\hat{\rho}\,\mathrm{sgn}\left(S\right)/\left(\hat{l}_1 + \hat{l}_2\right) \\
& - Sg_1\phi_1 l_1 a_1^{-1}u_{r,1} - Sg_2\phi_2 l_2 a_2^{-1}u_{r,2}.
\end{aligned} \tag{7.45}$$

By applying Eq. (7.35), one can achieve:

$$\begin{aligned}
& Sg_1\phi_1 l_1 K_1 / \left(\hat{l}_1 + \hat{l}_2\right) + Sg_2\phi_2 l_2 K_2 / \left(\hat{l}_1 + \hat{l}_2\right) \\
=\ & S\left[\ddot{\theta}_d - \eta\left(\dot{\theta} - \dot{\theta}_d\right) - f_2\left(x\right)\right]\left(l_1 + l_2\right)/\left(\hat{l}_1 + \hat{l}_2\right).
\end{aligned} \tag{7.46}$$

Recalling Assumption 7.4 yields:

$$-\frac{Sg_1\phi_1 l_1 g_1^{-1}\hat{\rho}\,\mathrm{sgn}\left(S\right)}{\hat{l}_1 + \hat{l}} - \frac{Sg_2\phi_2 l_2 g_2^{-1}\hat{\rho}\,\mathrm{sgn}\left(S\right)}{\hat{l}_1 + \hat{l}} \leq \frac{l_1 + l_2}{\hat{l}_1 + \hat{l}_2}\mu_1\hat{\rho}\,|S|. \tag{7.47}$$

In accordance with Assumption 7.5 and Assumption 7.6, the following inequalities exist:

$$-Sg_1\phi_1 l_1 a_1^{-1}\mu_{r,1} \leq l_1\varepsilon\,|S|\,\bar{u}_r \leq \varepsilon\,|S|\,\bar{u}_r, \tag{7.48}$$

$$-Sg_2\phi_2 l_2 a_2^{-1}\mu_{r,2} \leq l_2\varepsilon\,|S|\,\bar{u}_r \leq \varepsilon\,|S|\,\bar{u}_r. \tag{7.49}$$

Thus, from Eq. (7.46) and inequalities (7.47)–(7.49), the time derivative of V_1 can be represented as:

$$\begin{aligned}
\dot{V}_1 \leq\ & S\left[f_2\left(x\right) - \ddot{\theta}_d + \eta\left(\dot{\theta} - \dot{\theta}_d\right)\right] - \frac{l_1 + l_2}{\hat{l}_1 + \hat{l}_2}\mu_1\hat{\rho}\,|S| + 2\varepsilon\,|S|\,\bar{u}_r \\
& + \frac{l_1 + l_2}{\hat{l}_1 + \hat{l}_2}S\left[\ddot{\theta}_d - \eta\left(\dot{\theta} - \dot{\theta}_d\right) - f_2\left(x\right)\right].
\end{aligned} \tag{7.50}$$

Since $\underline{\delta} \leq \delta_1$, $\delta_2 \leq \bar{\delta}$, and $\delta_M = \max\left\{|\underline{\delta}|, |\bar{\delta}|\right\}$, the following inequality holds:

$$\dot{V}_1 \leq \delta_M\,|S|\,|f_2^*\left(x\right)| - \left(1 + \underline{\delta}\right)\mu_1\hat{\rho}\,|S| + 2\varepsilon\,|S|\,\bar{u}_r. \tag{7.51}$$

Step 2) Using $\dot{\rho}_0 = c_0 \mu_1 |S|$, the time derivation of V_2 is:

$$\dot{V}_2 \leq -\rho_0 \mu_1 |S| + (1 + \underline{\delta}) \hat{\rho}_0 \mu_1 |S|. \tag{7.52}$$

Step 3) Using $\dot{\rho}_1 = c_1 \mu_1 |S| |q|$, the time derivation of V_3 is:

$$\dot{V}_3 \leq -\rho_0 \mu_1 |S| |q| + (1 + \underline{\delta}) \hat{\rho}_0 \mu_1 |S| |q|. \tag{7.53}$$

Step 4) Using $\dot{\rho}_2 = c_2 \mu_1 |S| |f_2^*(x)|$, the time derivation of V_3 is:

$$\dot{V}_4 \leq -\rho_2 \mu_1 |S| |f_2^*(x)| + (1 + \underline{\delta}) \hat{\rho}_2 \mu_1 |S| |f_2^*(x)|. \tag{7.54}$$

Step 5) Using $\dot{\hat{\varepsilon}} = c_3 \mu_1 |S| \bar{u}_r$, the time derivation of V_5 is:

$$\dot{V}_5 \leq -2\varepsilon |S| \bar{u}_r + (1+) \hat{\varepsilon} \mu_1 |S| \bar{u}_r. \tag{7.55}$$

Step 6) Combining (7.52)–(7.55) and applying $\hat{\rho} = \hat{\rho}_0 + \hat{\rho}_1 |q| + \hat{\rho}_2 |f_2^*(x)| + \hat{\varepsilon}\bar{u}_r$ can achieve:

$$
\begin{aligned}
\dot{V}_2 + \dot{V}_3 + \dot{V}_4 + \dot{V}_5 \leq & -\rho_0 \mu_1 |S| - \rho_1 \mu_1 |S| |q| - \rho_2 \mu_1 |S| |f_2^*(x)| \\
& - 2\varepsilon |S| \bar{u} + (1 + \underline{\delta}) \mu_1 \hat{\rho} |S|.
\end{aligned} \tag{7.56}
$$

In addition, combining (7.51) and (7.56) gives:

$$\dot{V} \leq -(\rho_2 \mu_1 - \delta_M) |S| |f_2^*(x)| - \rho_0 \mu_1 |S| - \rho_1 \mu_1 |S| |q|. \tag{7.57}$$

The condition $\rho_2 \geq \delta_M / \mu_1$ ensures $\dot{V} \leq 0$, which is the standard reachability condition [185] and is sufficient to guarantee the sliding is maintained. $\square$

Remark 7.4. K_1 and K_2 in Eq. (7.35) are designed using the feedback linearization technique, which is detailed in Ref. [186]. The actual control commands with respect to the RE and LE are solely dependent on the estimated values $\hat{l}_1$ and $\hat{l}_2$. Nonetheless, the proposed law possesses the capability of tolerating the inaccurate estimation of the actuator effectiveness factors.

Remark 7.5. The amplitude and rate limits of flight actuators are included in Eq. (7.27), based on which the accommodation strategy is developed against actuator failures. The indicator $0 < \phi_i \leq 1$ is adopted to describe the amplitude saturation degree of the ith actuator. When the indicator ϕ_i approaches closer to 0, the more severe exhaustion is present. The adaptive laws outlined in Eqs. (7.37)–(7.40) are directly pertaining to the parameter μ_1 satisfying $0 < \mu_1 \leq \min_i (g_i \phi_i g_i^{-1})$. In regard to the actuation rate limit, the adaptive estimation of ε is related to the maximum actuation rate $(\bar{u}_r)$ as shown in Eq. (7.40). In summary, the control authority of the actuators is incorporated into the design procedure. One significant merit of the proposed methodology lies in respecting both the actuator amplitude and rate constraints.

Remark 7.6. *The number of design parameters in the developed scheme is more than that in the schemes without considering actuator authority. This fact arises due to that the constraints of both actuator amplitude and rate need to be handled in the proposed design. Note that c_3 is directly related to the convergence rate of the estimation of ε. The value of ε is adapted according to the actuator rate limit in the proposed algorithm. In addition, as inequality (7.57) indicates, the parameter μ_1 plays a dominant role in system performance since the parameter can affect the convergence rate and accuracy of the state variables by resorting to the sliding reaching law. In terms of Eq. (7.22) and Assumption 7.5, the parameter μ_1 is pertinent to the upper bound of the actuator amplitude. Thus, the larger $\bar{\mu}_i$ is, the bigger μ_1 becomes while a faster and more accurate response can be obtained. The similar statement can be also applied to the actuation rate bound.*

Remark 7.7. *A standard FCS attempts to control the rotational rates or attitudes using the control surface deflections to provide the right combination of roll, pitch, and yaw moments. The generation of a combined roll, pitch, and yaw demand requires a balanced combination of control surface deflections. This study focuses on the longitudinal channel of Boeing 747 aircraft. Hence, as far as the pitch angle tracking is only concerned, the RE and LE are control surfaces providing the appropriate pitching moment. For instance, when LE is outage, RE can be managed to counteract the detrimental effect of the faulty LE and complete the pitching maneuver. The unfavorable rolling moment induced by the RE can be compensated by the ailerons, which should be handled using the six degrees of freedom aircraft model. Further, with respect to passenger aircraft, RE can be constructed by right inner elevator and right outer elevator, while LE can be formed by left inner elevator and left outer elevator [181]. In this more complex case, control allocation techniques can also be applied to assign control signals to a set of surfaces [181, 187, 188].*

7.4   Fault Accommodation with Actuator Constraints and Sensorless Angular Rate

7.4.1   An Overview of the SMO-Based Fault Accommodation Scheme with Sensorless Angular Velocity

Distinctive from Chapter 7.3, this part goes a step further by considering the case where no measured pitch rate is available while actuator failures are present. From Fig. 7.3, the accommodation strategy aims at accommodating aircraft actuator failures without measured pitch rate, where the physical limits on healthy actuators are also respected. The system is constituted by an FDD unit, a SMO, and an adaptive SMC-based safety flight control, respec-

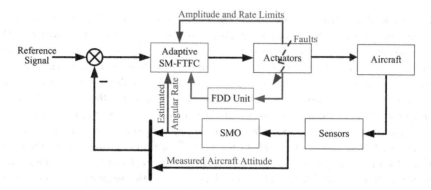

FIGURE 7.3: Illustrative diagram of the safety control scheme without measured angular rates.

tively. The SMO serves to reconstruct the information of aircraft pitch rate, which is ultimately applied to the presented scheme. Subsequently, the control is able to mitigate the negative effect caused by the faulty actuator.

Remark 7.8. *Within a sensor FDD mechanism, fault detection is performed by evaluating the residual between each sensor and its associated estimate based on the concept of analytical redundancy with several different methods in Refs. [189, 190, 191]. If the residual exceeds a specific threshold, the alarm for that sensor fault is triggered. Fault identification is then carried out by locating which sensor fault alarm is on. Once the failed sensor is identified, it remains in the failed state throughout the process. Moreover, the faulty sensor reading is replaced with a reliable estimate, which is a process known as sensor fault accommodation. The main focus is placed on the pitch rate reconstruction and the fault accommodation. The details of gyro fault detection and isolation are ignored due to the limited space.*

7.4.2   A SMO for Estimating Angular Rate

A SMO is developed as follows:

$$\begin{cases} \dot{\hat{\theta}} = \hat{q} + \tau_{01}\text{sgn}\left(\theta - \hat{\theta}\right) \\ \dot{\hat{q}} = f_2\left(\gamma, \theta, \hat{q}\right) + gL\phi\left(u_c - a^{-1}u_r\right) + \tau_{02}\text{sgn}\left(\theta - \hat{\theta}\right) \end{cases}, \qquad (7.58)$$

where $\hat{\theta}$ and $\hat{q}$ stand for the estimated values of θ and q, and τ_{01} and τ_{02} are positive scalars to be determined.

Remark 7.9. *The difficulty that the angular gyro no longer offers reliable angular rate is addressed, however the pitch angle θ can still be reliably measured. The estimated value of θ is only used for reconstructing $\hat{q}$ within the proposed SMO structure.*

Define the estimated errors with respect to θ and q as:

$$\begin{cases} \tilde{\theta} \overset{\Delta}{=} \theta - \hat{\theta} \\ \tilde{q} \overset{\Delta}{=} q - \hat{q} \end{cases} \tag{7.59}$$

Based on the aircraft longitudinal model in Eq. (7.27) and on the SMO representation in Eq. (7.58), the estimated error dynamics can be described as:

$$\begin{cases} \dot{\tilde{\theta}} = \tilde{q} - \tau_{01} \mathrm{sgn}\left(\tilde{\theta}\right) \\ \dot{\tilde{q}} = (f_2(x) - f_2(\gamma, \theta, \hat{q})) - \tau_{02} \mathrm{sgn}\left(\tilde{\theta}\right) \end{cases} \tag{7.60}$$

7.4.3 Integrated Design of SMO and Fault Accommodation

The actuator failures are tolerated by resorting to the healthy actuators within the proposed scheme. At the design stage, both the amplitude and rate limits of the healthy actuators are also accounted for. Further, the inaccurate estimation of l_1 and l_2 is incorporated. The SMO is adopted to recover the pitch rate information since the sensed pitch rate is no longer accessible. Instead of the strategy in Chapter 7.3, the estimated pitch rate $\hat{q}$ is utilized together with the measured pitch angle θ for the designed safety control loop.

In order to accomplish $\theta \to \theta_d$, the sliding surface similar to Eq. (7.31) is selected as:

$$S_{oc} = \hat{q} - \dot{\theta}_d + \eta_c (\theta - \theta_d), \tag{7.61}$$

where $\eta_c > 0$ is a design parameter.

Assumption 7.7. *There exist a positive constant χ and a compact set D_1 such that $D_1 = \left\{ \left(\theta, q, \hat{\theta}, \hat{q}\right) \middle| |\theta| \leq \chi, |q| \leq \chi, \left|\hat{\theta}\right| \leq \chi, |\hat{q}| \leq \chi \right\}$. In addition, it is also assumed that $|\theta_d| \leq \chi$ and $\left|\dot{\theta}_d\right| \leq \chi$.*

According to the detailed mathematical model of $\dot{q}$ [179], one obtains that $f_2(x) - f_2(\gamma, \theta, \hat{q}) = -\Xi \tilde{q}$, where Ξ is a positive bounded scalar. This condition as well as Assumption 7.7 can be adopted for the proof procedure of Theorem 7.2.

Remark 7.10. *With respect to the Boeing 747-100/200 aircraft, physical limits on the pitch angle and the corresponding angular rate are exposed for the sake of flight envelope protection. The estimated values of the pitch angle and angular rate are bounded as long as the SMO is properly designed. Moreover, the command signals in an FCS also have the constraints on the amplitude and rate for preventing dangerous maneuvers. Thus, Assumption 7.7 made for the convenience of the proof procedure of Theorem 7.2 is reasonable from a practical perspective. It is noted that the safety control aims at maintaining a satisfactory level of pitch tracking performance within the amplitude and* *rate limits of actuators. In the feedback control loop, only the information of pitch angle and pitch rate is required. Therefore, the limit on the FPA is not included in Assumption 7.7.*

Theorem 7.2. *Consider the aircraft longitudinal model subject to actuator faults and limited control authority in Eq. (7.27) and the observer constructed by Eq. (7.58), where the observer design parameters τ_{01} and τ_{02} satisfying:*

$$\begin{cases} \tau_{01} \geq 2\chi \\ \tau_{02} \leq 2\Xi\chi/\left(2 + \eta_c\right) \end{cases}. \tag{7.62}$$

There exists a safety control law that is determined as:

$$\begin{cases} u_{c,1} = \left[K_1 - g_1^{-1}\hat{\rho}\mathrm{sgn}\left(S_{oc}\right)\right]/\left(\hat{l}_1 + \hat{l}_2\right) \\ u_{c,2} = \left[K_2 - g_2^{-1}\hat{\rho}\mathrm{sgn}\left(S_{oc}\right)\right]/\left(\hat{l}_1 + \hat{l}_2\right) \end{cases}, \tag{7.63}$$

where

$$\begin{cases} K_1 = \phi_1^{-1}g_1^{-1}\left[\ddot{\theta}_d - \eta_c\left(\hat{q} - \dot{\theta}_d\right) - f_2\left(\gamma, \theta, \hat{q}\right)\right] \\ K_2 = \phi_2^{-1}g_2^{-1}\left[\ddot{\theta}_d - \eta_c\left(\hat{q} - \dot{\theta}_d\right) - f_2\left(\gamma, \theta, \hat{q}\right)\right] \end{cases}. \tag{7.64}$$

Further, the estimation of the modulation gain is:

$$\hat{\rho} = \hat{\rho}_0 + \hat{\rho}_1 |q| + \hat{\rho}_2 |f_2^*\left(x\right)| + \hat{\varepsilon}\bar{u}_r, \tag{7.65}$$

and the update laws are:

$$\dot{\hat{\rho}}_0 = c_0\mu_1 |S_{oc}|, \tag{7.66}$$

$$\dot{\hat{\rho}}_1 = c_1\mu_1 |S_{oc}| |\hat{q}|, \tag{7.67}$$

$$\dot{\hat{\rho}}_2 = c_2\mu_1 |S_{oc}| |f_2^*\left(x\right)|, \tag{7.68}$$

$$\dot{\hat{\varepsilon}} = c_3\mu_1 |S_{oc}| \bar{u}_r, \tag{7.69}$$

where $\hat{\rho}$, $\hat{\rho}_0$, $\hat{\rho}_1$, $\hat{\rho}_2$, and $\hat{\varepsilon}$ denote the estimated values of the positive constants ρ, ρ_0, ρ_1, ρ_2, and ε and $\rho_2 \geq \delta_M/\mu_1$, $c_j > 0 \, (j = 0, \cdots, 3)$ are design parameters, and $f_2^*\left(x\right) = \ddot{\theta}_d - \eta\left(\hat{q} - \dot{\theta}_d\right) - f_2\left(\gamma, \theta, \hat{q}\right)$. Then the trajectory of the faulty closed-loop system can be driven onto the sliding surface $S_{oc} = 0$ and it can eventually converge to the origin. In this case, all of the states of the closed-loop system (i.e., θ and q) can be stabilized under actuator failures.

Proof. Choose a Lyapunov candidate as:

$$\begin{aligned} V_{oc} = {} & S_{oc}^2/2 + (\rho_0 - (1+\underline{\delta})\hat{\rho}_0)^2/(2(1+\underline{\delta})c_0) \\ & + (\rho_1 - (1+\underline{\delta})\hat{\rho}_1)^2/(2(1+\underline{\delta})c_1) \\ & + (\rho_2 - (1+\underline{\delta})\hat{\rho}_2)^2/(2(1+\underline{\delta})c_2) \\ & + (2\varepsilon/\mu_1 - (1+)\hat{\varepsilon})^2/(2(1+\underline{\delta})c_3) \\ & + \tilde{\theta}^2/2 + \tilde{q}^2/2. \end{aligned} \tag{7.70}$$

From the aspects of safety control and observer, the Lyapunov function can be split into two portions as following:

$$V_{oc} = V_c + V_o, \tag{7.71}$$

where

$$
\begin{aligned}
V_c =\ & S_{oc}^2/2 + (\rho_0 - (1+\underline{\delta})\hat{\rho}_0)^2/(2(1+\underline{\delta})c_0) \\
& + (\rho_1 - (1+\underline{\delta})\hat{\rho}_1)^2/(2(1+\underline{\delta})c_1) \\
& + (\rho_2 - (1+\underline{\delta})\hat{\rho}_2)^2/(2(1+\underline{\delta})c_2) \\
& + (2\varepsilon/\mu_1 - (1+)\hat{\varepsilon})^2/(2(1+\underline{\delta})c_3),
\end{aligned}
\tag{7.72}
$$

and

$$V_o = \tilde{\theta}^2/2 + \tilde{q}^2/2. \tag{7.73}$$

The proof procedure is conducted by three steps: (1) the first step is to analyze the time derivative of V_c; (2) the second one is to yield $\dot{V}_o$; and (3) the final step is to combine $\dot{V}_o$ and $\dot{V}_c$.

Step 1) The time derivative of the sliding surface S_{oc} is:

$$\dot{S}_{oc} = \dot{\hat{q}} - \ddot{\theta}_d + \eta_c \left(\hat{q} - \dot{\theta}_d \right). \tag{7.74}$$

Applying the observer representation Eq. (7.58) and the safety control law Eq. (7.63) obtains:

$$
\begin{aligned}
\dot{S}_{oc} =\ & \left[f_2\left(\gamma, \theta, \hat{q}\right) - \ddot{\theta}_d + \eta_c \left(\hat{q} - \dot{\theta}_d\right) \right] \\
& + g_1\phi_1 l_1 K_1/\left(\hat{l}_1 + \hat{l}_2\right) + g_2\phi_2 l_2 K_2/\left(\hat{l}_1 + \hat{l}_2\right) \\
& - g_1\phi_1 l_1 g_1^{-1}\hat{\rho}\mathrm{sgn}\left(S_{oc}\right)/\left(\hat{l}_1 + \hat{l}_2\right) \\
& - g_1\phi_1 l_1 g_1^{-1}\hat{\rho}\mathrm{sgn}\left(S_{oc}\right)/\left(\hat{l}_1 + \hat{l}_2\right) \\
& - g_1\phi_1 l_1 a_1^{-1}u_{r,1} - g_2\phi_2 l_2 a_2^{-1}u_{r,2} + \tau_{02}\mathrm{sgn}\left(\tilde{\theta}\right).
\end{aligned}
\tag{7.75}
$$

Now, the time derivative of V_c can be written as:

$$
\begin{aligned}
\dot{V}_c =\ & S_{oc}\dot{S}_{oc} + (\rho_0 - (1+\underline{\delta})\hat{\rho}_0)\dot{\hat{\rho}}_0/c_0 \\
& - (\rho_1 - (1+\underline{\delta})\hat{\rho}_1)\dot{\hat{\rho}}_1/c_1 - (\rho_2 - (1+\underline{\delta})\hat{\rho}_2)\dot{\hat{\rho}}_2/c_2 \\
& - (2\varepsilon/\mu_1 - (1+\underline{\delta})\hat{\varepsilon})\dot{\hat{\varepsilon}}/c_3.
\end{aligned}
\tag{7.76}
$$

The detailed procedure is akin to the proof of Theorem 7.1. Applying Eqs. (7.66)–(7.69) and letting $f_2^*(x) = \ddot{\theta}_d - \eta\left(\hat{q} - \dot{\theta}_d\right) - f_2\left(\gamma, \theta, \hat{q}\right)$ can lead to:

$$
\begin{aligned}
\dot{V}_c \leq\ & - (\rho_2\mu_1 - \delta_M)\,|S_{oc}|\,|f_2^*(x)| - \rho_0\mu_1\,|S_{oc}| \\
& - \rho_1\mu_1\,|S_{oc}|\,|\hat{q}| + S_{oc}\tau_{02}\mathrm{sgn}\left(\tilde{\theta}\right).
\end{aligned}
\tag{7.77}
$$

Step 2) Applying the estimated error dynamics in Eq. (7.60), the time derivation of V_o is:

$$
\begin{aligned}
\dot{V}_o &= \tilde{\theta}\dot{\tilde{\theta}} + \tilde{q}\dot{\tilde{q}} \\
&= \tilde{\theta}\left(\tilde{q} - \tau_{01}\mathrm{sgn}\left(\tilde{\theta}\right)\right) \\
&\quad + \tilde{q}\left(f_2\left(x\right) - f_2\left(\gamma, \theta, \hat{q}\right) - \tau_{o2}\mathrm{sgn}\left(\tilde{\theta}\right)\right) \\
&\leq \left|\tilde{\theta}\right|\left|\tilde{q}\right| - \tau_{01}\left|\tilde{\theta}\right| - \Xi\tilde{q}^2 + \tau_{o2}\left|\tilde{q}\right|.
\end{aligned}
\tag{7.78}
$$

Step 3) Recalling Assumption 7.7 and then combining (7.77) and (7.78) can result in:

$$
\begin{aligned}
\dot{V}_{oc} &= \dot{V}_c + \dot{V}_o \\
&\leq -\left(\rho_2\mu_1 - \delta_M\right)\left|S_{oc}\right|\left|f_2^*\left(x\right)\right| \\
&\quad - \rho_0\mu_1\left|S_{oc}\right| - \rho_1\mu_1\left|S_{oc}\right|\left|\hat{q}\right| - \tilde{\theta}\left(\tau_{01} - \left|\tilde{q}\right|\right) \\
&\quad - \left(\Xi\tilde{q}^2 - \tau_{o2}\left|\tilde{q}\right| - \tau_{o2}\left|S_{oc}\right|\right).
\end{aligned}
\tag{7.79}
$$

When the design parameter τ_{01} is selected as $\tau_{01} \geq 2\chi$, the fourth term of the right side of (7.79) $-\left|\tilde{\theta}\right|\left(\tau_{01} - \left|\tilde{q}\right|\right) \leq 0$ is ensured. If the design parameter τ_{02} is chosen as $\tau_{02} \leq 2\Xi\chi\left(2+\eta_c\right)$, the fifth term of the right side of (7.79) $-\left(\Xi\tilde{q}^2 - \tau_{o2}\left|\tilde{q}\right| - \tau_{o2}\left|S_{oc}\right|\right) \leq 0$ holds. Further, the condition $\rho_2 \geq \delta_M/\mu_1$ can guarantee $\dot{V}_{oc} < 0$, which indicates that all of the states of the closed-loop system (i.e., θ and q) can be stabilized under actuator failures.   □

Remark 7.11. *It is worth mentioning that the separation principle can facilitate the individual design of observer and control for linear systems. In contrast, this principle is not fitted for nonlinear systems [192]. Thus, the integrated design of SMO and safety control is investigated for the nonlinear longitudinal aircraft model. The combination of the SMO and accommodation allows the system to compensate for actuator faults without measured pitch rate. The safety control synthesis is interacted with the SMO design. It is obvious from Ref. (7.62) that the determination of the observer design parameter τ_{02} is tightly relevant to the design parameter η_c of the safety control.*

Remark 7.12. *In the absence of measured angular rate, one option to remedy the angular rate feedback is to produce a velocity signal by numerically differentiating the sensed attitude. The quantization effect may induce unfavorable oscillations and deteriorate the stability of the closed-loop system. The existing FDD design to deal with aircraft sensor faults includes observer based Refs. [14, 174, 175] and neural network based Ref. [176] methods. The significant difference between the existing method and the proposed one is that the severe situation of both actuator faults and angular rate gyro faults is explicitly considered in this chapter. To be more specific, a SMO is exploited to reconstruct the pitch rate once the angular rate gyro onboard is not reliable.*

Subsequently, the recovered pitch rate and the measured pitch angle are used for the safety control loop, thereby preserving the tracking performance. Therefore, the proposed methodology can be regarded as a backup plan in urgent and severe cases where the gyro sensor cannot serve any more and the elevator actuators fail.

7.5 Simulation Studies

7.5.1 Simulation Environment Description

The trimming condition of the Boeing 747 transport aircraft is indicated in Table 7.1, while the control authority of the RE and LE is presented in Table 7.2. The sampling interval for the Boeing aircraft control is chosen as 0.01 s.

The reference signal is generated after passing through the following pre-filter:

$$\ddot{\theta}_d + 3\dot{\theta}_d + 4\theta_d = 4\theta^*, \tag{7.80}$$

where the original command θ^* is set with the magnitude of $5.74°$ and the duration of 15 s.

7.5.2 Simulation Scenarios

The simulation scenarios are briefly summarized in Table 7.3. The LE outage ($l_2=0$) is invoked at 13 s of the simulation. The amount of time consumed by the FDD unit is assumed to be 1 s such that reliable FDD results can be achieved for control reconfiguration. The sensor gyro for measuring the pitch rate is available in Case I, while the sensorless pitch rate case is examined in

TABLE 7.1: Trimming values of the Benchmark Aircraft.

Symbol	Physical meaning	Value
H^{trim}	Altitude at the trimming condition	7000 m
V^{trim}	Airspeed at the trimming condition	230 m/s
α^{trim}	AOA at the trimming condition	0.9283°
θ^{trim}	Pitch angle at the trimming condition	0.9283°
q^{trim}	Pitch rate at the trimming condition	0
T^{trim}	Engine thrust at the trimming condition	41 631 N
δ_{re}^{trim}	RE deflection at the trimming condition	0
δ_{le}^{trim}	LE deflection at the trimming condition	0
δ_{hs}^{trim}	Horizontal stabilizer deflection at the trimming condition	0.7334°

TABLE 7.2: The operating limits of the elevators.

Control surface	Symbol	Amplitude limit	Rate limit
Right elevator	δ_{re}	$[-20°,20°]$	$[-37°/\text{s},37°/\text{s}]$
Left elevator	δ_{le}	$[-20°,20°]$	$[-37°/\text{s},37°/\text{s}]$

TABLE 7.3: The simulation scenarios.

	Case I-A	Case I-B	Case II-A	Case II-B
The LE is outage	√	√	√	√
q can be measured by the gyro	√	√		
q is reconstructed by SMO			√	√
FDD time is 1 s	√	√	√	√
Model uncertainty	√	√	√	√
FDD inaccuracy	√	√	√	√
Noises in measurements		√		√
Input delay		√		√

Case II. Further, in terms of the scenarios set by Refs. [171, 191, 193], the phenomena including model uncertainties, noise measurements, input delay, and inaccurate estimation of actuator effectiveness factors are also introduced in the two scenarios. There exists 20% mismatch in the mass of the aircraft, and also in the aerodynamic coefficients (C_L and C_m). The noise with a mean of 0 and covariance of 0.01 is injected into each measurement channel. The input delay is fixed 0.01 s as well as the sampling interval. The estimation of the RE effectiveness indicator is 90% of its true value, while the estimated value of the LE effectiveness indicator is 50% of its true value.

Three types of fault accommodation architectures are compared in the simulations. They are: (1) the SMC-based fault accommodation without accounting for actuator physical constraints; (2) the adaptive SMC-based fault with consideration of actuator amplitude bounds; and (3) the developed accommodation scheme with consideration of both actuator amplitude and rate limits. In the interest of brevity, the three schemes are named as (1) FTFC, (2) FTFC-AL, and (3) FTFC-ARL, respectively. The following metric is defined to quantitatively compare the selected schemes,

$$e_{perf} = \sqrt{\int_{t_1}^{t_2} \left[(\theta - \theta_d)^2 + \delta_{re}^2 + \dot{\delta}_{re}^2 \right] dt / (t_2 - t_1)}, \qquad (7.81)$$

where $[t_1,t_2]$ covers the simulation duration. The index in Eq. (7.81) describes the performance over the simulation interval considering both tracking error and actuation movements.

7.5.3 Results of Case I and Assessment

(1) Case I-A

As depicted in Fig. 7.4, in the event of LE outage, the pitch angle response becomes oscillatory when the reference input tends to the trimming value (0.9283°) under the FTFC scheme. In contrast, the FTFC-AL and FTFC-ARL can preserve an acceptable level of pitch tracking performance. Fig. 7.5 reveals the resulting deflections of the elevators. The RE governed by the FTFC reaches the magnitude limits (20°) during the transient process. However, the RE can stay within the amplitude bounds to compensate for the impact due to the LE damage, when the FTFC-AL and FTFC-ARL are commissioned. Key observations of Fig. 7.6 include that: (1) the RE deflection rate saturation is encountered in the presence of both the FTFC and FTFC-AL schemes; and (2) the RE deflection rate under the FTFC-ARL is within the allowable range ($[-37°/s, 37°/s]$). It should be mentioned that the pitch rate of the FTFC-ARL is below those of the FTFC and FTFC-AL. This response coincides with the fact that the RE reacts less aggressively to avoid the control authority exhaustion.

The quantitative performance indices are included in Table 7.4. The peak values of the RE deflection are 20°, 10.22°, and 6.47°, corresponding to the FTFC, FTFC-AL, and FTFC-ARL, respectively. The RE rate exhaustion (37°/s) is present under the FTFC and FTFC-AL systems. On the other hand, the peak value of the RE rate is 23.87°/s in the case of FTFC-ARL. The FTFC-ARL also outperforms the FTFC and FTFC-AL in terms of the index e_{perf}. The enhanced rate from the FTFC to the FTFC-ARL is 85.78%

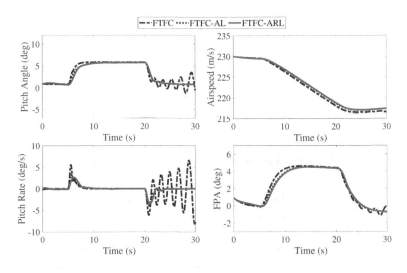

FIGURE 7.4: The responses of aircraft states in Case I-A.

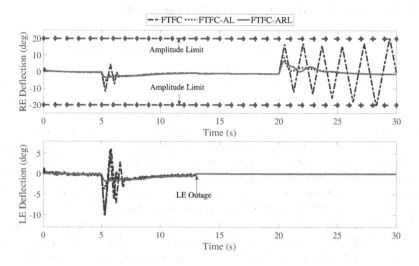

FIGURE 7.5: The deflections of elevators in Case I-A.

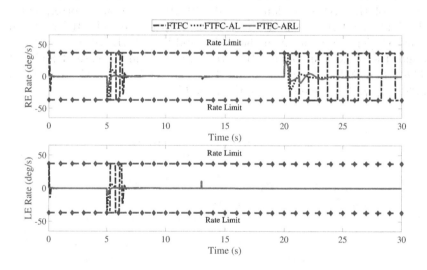

FIGURE 7.6: The deflection rates of elevators in Case I-A.

(from 44.79 to 6.37), while the improved percentage from the FTFC-AL to the FTFC-ARL is 40.36% (from 10.68 to 6.37).

(2) Case I-B

Figs. 7.7–7.9 illustrate that the FTFC-ARL scheme can achieve superior performance than those under the FTFC and FTFC-AL schemes, when the measurement noises and input delay are taken into account. To be more specific, the indices listed in Table 7.4 confirm that the LE deflection and deflec-

TABLE 7.4: Performance comparison in Case I.

		FTFC	FTFC-AL	FTFC-ARL
Case I-A	$\left|\delta_{re}\right|_{max}(\circ)$	20	10.22	6.47
	$\left|\dot{\delta}_{re}\right|_{max}(\circ/s)$	37	37	23.87
	e_{perf}	44.79	10.68	6.37
Case I-B	$\left|\delta_{re}\right|_{max}(\circ)$	20	10.85	6.61
	$\left|\dot{\delta}_{re}\right|_{max}(\circ/s)$	37	37	35.14
	e_{perf}	60.36	23.86	18.97

tion rate are limited to the allowable ranges when the FTFC-ARL is engaged. The defined index achieved by the FTFC-ARL is better than those under the FTFC and FTFC-AL. Hence, in spite of more realistic factors, the presented FTFC-ARL system can achieve superior performance than that under the FTFC and FTFC-AL schemes.

7.5.4 Results of Case II and Assessment

(1) Case II-A

It is highlighted in Fig. 7.10 that the FTFC-AL and FTFC-ARL ensure that the pitch angle can track the reference input after the LE failure. Instead, the FTFC scheme results in the pitch angle excursions. The RE managed by the FTFC-AL and FTFC-ARL is operating within the travel limits, as

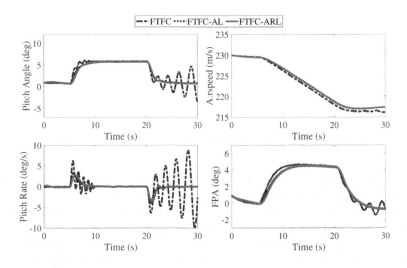

FIGURE 7.7: The responses of aircraft states in Case I-B.

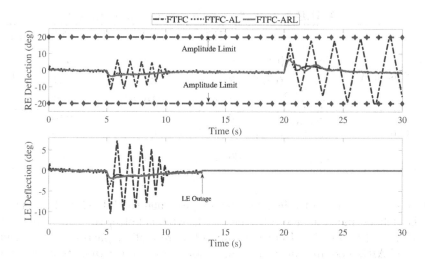

FIGURE 7.8: The deflections of elevators in Case I-B.

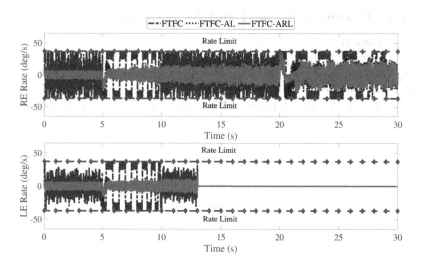

FIGURE 7.9: The deflection rates of elevators in Case I-B.

evidenced in Fig. 7.11. The RE deflection of the FTFC-ARL is smoother in comparison with those of the FTFC-AL and FTFC. What is interesting to see further is that the pitch angle tracking of the FTFC-ARL is slightly slower than the tracking of the FTFC-AL, thereby avoiding the exhaustive usage of RE. Fig. 7.12 exhibits the attractive benefit of the FTFC-ARL, not exceeding the actuator rate limits, in addition to keeping the flight safety. In this sense, the performance of the FTFC and FTFC-AL is inferior to that of

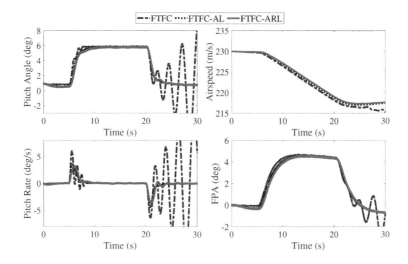

FIGURE 7.10: The responses of aircraft states in Case II-A.

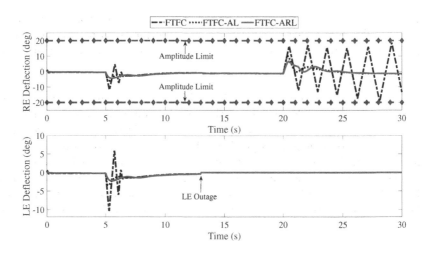

FIGURE 7.11: The deflections of elevators in Case II-A.

the proposed FTFC-ARL. Finally, as can be seen in Fig. 7.13, the proposed SMO is capable of recovering the pitch rate information, which is reliable for the FTFC purpose.

Table 7.5 indicates that the RE with the peak deflections of $10.24°$ and $6.66°$ is managed to counteract the LE outage under the FTFC-AL and FTFC-ARL, respectively. The RE actuation rates continuously touch the physical bounds in the event of the FTFC and FTFC-AL. However, the peak deflection rate of the RE is $24.31°$/s under the control of the developed FTFC-

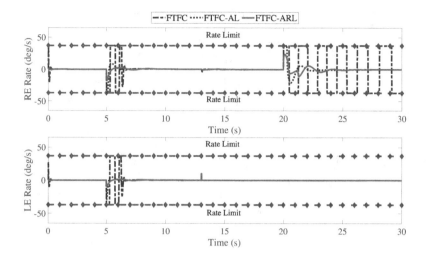

FIGURE 7.12: The deflection rates of elevators in Case II-A.

ARL. The defined index e_{perf} is significantly improved by 85.19% (from 44.90 to 6.65). The FTFC-ARL also attains the superior performance compared to the FTFC-AL, with 38.48% improvement of the defined metric (from 10.81 to 6.65).

(2) Case II-B

In addition to Case II-A, the noise is injected into each measurement channel, while the input delay is fixed as the sampling interval. Figs. 7.14–7.17 show that the FTFC-ARL scheme can guarantee the flight safety in the event of LE outage and unreliable gyroscope. The FTFC-ARL can also ensure that the RE works within the physical limits during the fault accommodation process. The indices in Table V exemplify that the developed FTFC scheme outperforms the FTFC and FTFC-AL schemes in the realistic simulation environment. It should be noticed that the performance of Case II-B achieved by the designed

TABLE 7.5: Performance comparison in Case II.

		FTFC	FTFC-AL	FTFC-ARL
Case II-A	$\|\delta_{re}\|_{max}$ (∘)	20	10.24	6.66
	$\|\dot{\delta}_{re}\|_{max}$ (∘/s)	37	37	24.31
	e_{perf}	44.90	10.81	6.65
Case II-B	$\|\delta_{re}\|_{max}$ (∘)	20	11.05	6.72
	$\|\dot{\delta}_{re}\|_{max}$ (∘/s)	37	37	35.26
	e_{perf}	60.56	26.67	19.34

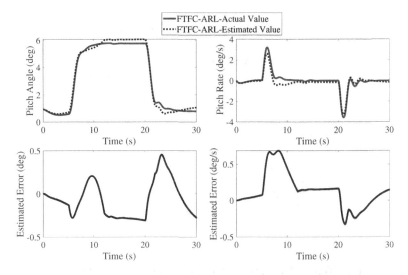

FIGURE 7.13: The actual (solid line) and reconstructed (dashed line) pitch angle (top left) and pitch rate (top right) in the proposed scheme, and the estimated errors of pitch angle (bottom left) and pitch rate (bottom right).

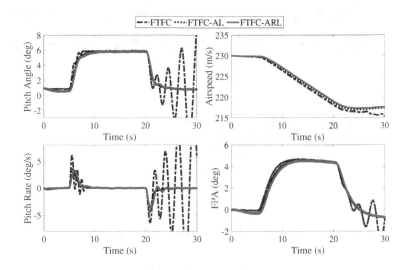

FIGURE 7.14: The responses of aircraft states in Case II-B.

scheme is slightly degraded than that of Case II-A, since the measurement noises and input delay are involved.

Finally, it is worth emphasizing that although the Boeing 747 transport aircraft in longitudinal motion has been selected to demonstrate the concept

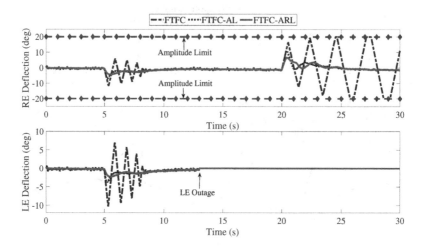

FIGURE 7.15: The deflections of elevators in Case II-B.

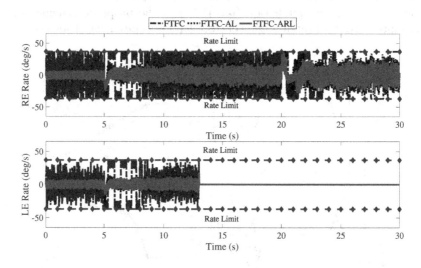

FIGURE 7.16: The deflection rates of elevators in Case II-B.

and effectiveness, the proposed design strategies are general and applicable for different types of aircraft.

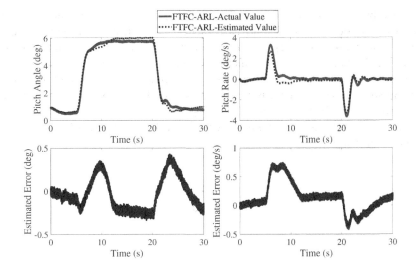

FIGURE 7.17: The actual (solid line) and reconstructed (dashed line) pitch angle (top left) and pitch rate (top right) in the proposed scheme, and the estimated errors of pitch angle (bottom left) and pitch rate (bottom right).

7.6 Conclusions

Unexpected magnitude and rate limiting in healthy flight actuators can reduce the stability margin, and even deteriorate the safety of the post-failure aircraft. A new safety flight control scheme against actuator failures is proposed by incorporating both the actuator amplitude and rate bounds. Further, potential sensor fault on the measured angular rate is considered in the design procedure. In the absence of a reliable gyro onboard, the sensed pitch rate is delivered into the safety control loop. In the presence of gyro sensor fault, which leads to the sensed angular rate becoming no longer accessible, the pitch rate information is reconstructed by a SMO for the safety control feedback.

7.7 Notes

Compared to the conventional SMC, an advantage of the proposed sliding surface with integral action is that the system trajectory always starts from the sliding surface. Through the numerical simulations, we can find that the unique advantages of the proposed methodologies include that: under actuator failures, (1) the magnitude and rate saturation in healthy actuators can be notably prevented; and (2) the flight safety and tracking performance can be maintained whether or not the measured angular velocity is offered.

Despite that the proposed strategies are capable of tolerating flight actuator and sensor gyro failures, issues including modeling and sensing uncertainties have not yet been considered in the design. Investigation of these factors which may affect the performance of the overall safety control system is one of our future works.

Appendix A

Appendix for Chapter 2

As can be seen from Fig. A.1, the purpose of the actuator in a flight control system is to convert the control command u_{ci} to the actual deflection of the control surface u_i. The hydraulic pump provides the necessary pressure to drive the control valve, which transfers the desired control command to the pressure acted on the piston. The load on the control surface encountered during a flight also generates pressure to the piston in an opposite direction. The pressure differential results in a net force which moves the piston, therefore, the control surface. When all the torques generated by the hydraulic force balances that generated by the load, the piston will rest at a deflection angle proportional to the control command.

The nonlinear state equations relating the position of the valve spool D_{spool} to the control signal u_{ci} can be represented as [70]:

$$\dot{D}_{spool} = v_{spool}. \tag{A.1}$$

$$\dot{v}_{spool} = -\omega_v^2 D_{spool} - 2\xi_v \omega_v v_{spool} + k_v \omega_v^2 u_{ci}. \tag{A.2}$$

Revisiting Fig. A.1 with Eq. (A.2), the block diagram of valve from the control signal to the position of the valve spool consists of an integral part. The position of the valve spool and the spool speed are also used as feedback signals. Hence, there is an integral action internal to the valve. The terminal characteristics can be represented as a second order system Eqs. (A.1) and (A.2).

Based on the continuity equation of flow, the relationship among the load pressure P, rod displacement D_{rod}, and valve spool displacement D_{spool} can be written as,

$$k_1 D_{spool} = F\dot{D}_{rod} + (C_1 + C_2) P + \frac{1}{2}\left(k_e + \frac{V_0}{E}\right)\dot{P}, \tag{A.3}$$

where $k_1 = C_v w \sqrt{\frac{P_s - P}{\rho}}$.

The left hand side of Eq. (A.3) represents the output flow of the valve, the first term on right hand side of Eq. (A.3) is the flow due to piston motion, the second term stands for the flow due to piston seal leakage, and the third term is the flow that is resulted from variations in density and piston volume.

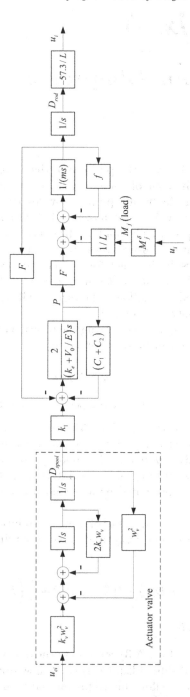

FIGURE A.1: Block diagram of a hydraulic driven actuator for control surface movement.

The pressure provided by the hydraulic equipment balances the external load to drive the actuator rod, the relationship can be described as [69, 70, 71, 194],

$$m\ddot{D}_{rod} = FP - f\dot{D}_{rod} - \frac{M_j}{L}, \tag{A.4}$$

where F is the piston effective area, f stands for piston damping coefficient, D_{rod} is the rod displacement, M_j is the external load, L is the length of rocker, and m is denoted as rod mass.

The relationship between the deflection of the control surface u_i and the rod displacement D_{rod} can be related as

$$u_i = -\frac{57.3}{L} D_{rod}. \tag{A.5}$$

The external load acted on the actuator during a flight can be expressed as

$$M_j = M_j^{\delta} \left(-\frac{57.3}{L} D_{rod} \right), \tag{A.6}$$

where M_j^{δ} is the hinge moment coefficient, which is closely related to airspeed, dynamic pressure, and altitude, can be obtained by wind-tunnel and flight tests.

After linearization, the equations become:

$$\dot{D}_{spool} = \Delta v_{spool}, \tag{A.7}$$

$$\dot{v}_{spool} = -\omega_v^2 \Delta D_{spool} - 2\xi_v \omega_v \Delta v_{spool} + k_v \omega_v^2 \Delta u, \tag{A.8}$$

$$k_1 \Delta D_{spool} = F\frac{d\Delta D_{rod}}{dt} + (C_1 + C_2)\Delta P + \frac{1}{2}\left(k_e + \frac{V_0}{E}\right)\dot{P}, \tag{A.9}$$

$$m\frac{d^2 \Delta D_{rod}}{dt^2} = F\Delta P - f\frac{d\Delta D_{rod}}{dt} - \frac{\Delta M_j}{L}, \tag{A.10}$$

$$\Delta M_j = M_j^{\delta}\left(-\frac{57.3}{L}\Delta D_{rod}\right), \tag{A.11}$$

where $\Delta M_j = 0$ (at force equilibrium point), the transfer function between D_{rod} and D_{spool} is

$$\frac{D_{rod}(s)}{D_{spool}(s)} = \frac{k_M}{s\left(T_M^2 s^2 + 2\xi_M T_M s + 1\right)}, \tag{A.12}$$

where $k_M = \frac{k_1 F}{F^2 + f(C_1 + C_2)}$, $T_M^2 = \frac{m(k_e + V_0/E)}{2[F^2 + f(C_1 + C_2)]}$, $\xi_M = \frac{2m(C_1 + C_2) + f(k_e + V_0/E)}{4T_M[F^2 + f(C_1 + C_2)]}$.

When $\Delta M_j \neq 0$ (load exerted on the actuator), the transfer function is

$$\frac{D_{rod}(s)}{D_{spool}(s)} = \frac{k_M}{T_M^2 s^3 + 2\xi_M T_M s^2 + \left[1 - \frac{57.3 k_M M_j^{\delta}}{2k_1 F L^2}\left(k_e + \frac{V_0}{E}\right)\right]s - \frac{57.3 k_M M_j^{\delta}(C_1 + C_2)}{k_1 F L^2}}. \tag{A.13}$$

Compared the time constant T_M with the that of an aircraft, the order of magnitude of T_M is nearly 10^{-3} sec, which is much smaller than time constant of an aircraft (e.g., in an F-16 fighter at 152 m/s, short-period mode T_s=4.21 sec, phugoid mode T_p=84.9 sec) [73]. Therefore, the time constant T_M can be reasonably ignored. Subsequently, Eq. (A.13) can be simplified to

$$\frac{D_{rod}(s)}{D_{spool}(s)} = \frac{k_M}{\left[1 - \frac{57.3k_M M_j^\delta}{2k_1 FL^2}\left(k_e + \frac{V_0}{E}\right)\right]s - \frac{57.3k_M M_j^\delta(C_1+C_2)}{k_1 FL^2}}. \qquad (A.14)$$

From Eqs. (A.1) and (A.2), the transfer function that relates the control signal u_{ci} to the position of the spool valve D_{spool} can be obtained,

$$\frac{D_{spool}(s)}{u_{ci}(s)} = \frac{k_v \omega_v^2}{s^2 + 2\xi_v \omega_v s + \omega_v^2}. \qquad (A.15)$$

The transfer function between the displacement of rod and deflection of control surface is

$$\frac{u_i(s)}{D_{rod}(s)} = \frac{-57.3}{L}. \qquad (A.16)$$

During a flight, the load is the force exerted to the control surface. Therefore, the transfer function between the actual deflection of the control surface and the control signal from the controller is

$$\frac{u_i(s)}{u_{ci}(s)} = \frac{k_M k_v \omega_v^2}{\left(\left(\frac{k_M M_j^\delta}{2k_1 FL}\left(k_e + \frac{V_0}{E}\right) - \frac{L}{57.3}\right)s + \frac{k_M M_j^\delta(C_1+C_2)}{k_1 FL}\right) \cdot (s^2 + 2\xi_v \omega_v s + \omega_v^2)}. \qquad (A.17)$$

Finally, Eq. (A.17) can be simplified as

$$\frac{u_i(s)}{u_{ci}(s)} = \frac{k_{ha} k_v}{(T_{ha}s + 1) \cdot \left(\frac{s^2}{\omega_v^2} + 2\frac{\xi_v}{\omega_v}s + 1\right)}, \qquad (A.18)$$

where $k_{ha} = \frac{k_1 FL}{M_j^\delta(C_1+C_2)}$ and $T_{ha} = \frac{k_M M_j^\delta\left(k_e + \frac{V_0}{E}\right) - \frac{2}{57.3}k_1 FL^2}{k_M M_j^\delta(C_1+C_2)}$.

Appendix B

Appendix for Chapter 3: Part 1

TABLE B.1: Symbols in Chapter 3.

Symbols	Explanations
u, v, w	Components of aircraft velocity along X, Y, and Z axes (m/s)
p, q, r	Aircraft roll rate, pitch rate, and yaw rate about body axis (rad/s)
p_w, q_w, r_w	Aircraft roll rate, pitch rate, and yaw rate about wind axis (rad/s)
α, β	Angle of attack, angle of sideslip (rad)
V, h	Total velocity (m/s), and flight altitude (m)
m	Aircraft total mass (kg)
M_a	Mach number
$\bar{q}$	Free stream dynamic pressure (N/m^2)
S	Wing surface (m^2)
b_{ref}, c_{ref}	Wingspan, and mean aerodynamic chord (m)
I_x, I_y, I_z	Moment of inertia about X, Y, and Z axes (kg $\cdot$ m^2)
I_{xz}	Product of inertia with respect to X and Z body axes (kg $\cdot$ m^2)
x_{cg}, y_{cg}, z_{cg}	Center of gravity location along X, Y, and Z axes
C_L, C_D, C_Y	Lift coefficient, drag coefficient, and side force coefficient
$C_{l_{tot}}, C_{m_{tot}}, C_{n_{tot}}$	Total rolling, pitching, and yawing-moment coefficient
D, L, Y	Drag, lift, and side-force in wind-fixed X, Y, and Z axes (N)
δ_{lc}, δ_{rc}	Left and right canard deflection (rad)
$\delta_{lie}, \delta_{rie}$	Left and right inner elevon deflection (rad)
$\delta_{loe}, \delta_{roe}$	Left and right outer elevon deflection (rad)
δ_r	Rudder deflection (rad)

$$C_1 = \left((I_y - I_z)I_z - I_{xz}^2\right)/\Gamma, \; C_2 = (I_x - I_y + I_z)I_{xz}/\Gamma, \; C_3 = I_z/\Gamma.$$

$$C_4 = I_{xz}/\Gamma, \; C_5 = I_z - I_x/I_y, \; C_6 = I_{xz}/I_y.$$

Continue to the table on the next page ......

Table B.1 (continued)

$$C_7 = 1/I_y \ C_8 = \left(I_x \left(I_x - I_y\right) + I_{xz}^2\right)/\Gamma, \ C_9 = I_x/\Gamma.$$

$$L = \bar{q} \cdot S \cdot C_L, D = \bar{q} \cdot S \cdot C_D, Y = \bar{q} \cdot S \cdot C_Y.$$

$$\Gamma = I_x I_z - I_{xz} I_{xz}, M_x = \bar{q} \cdot S \cdot b_{ref} \cdot C_{l_{tot}} - z_{cg} \cdot F_y + y_{cg} \cdot F_z.$$

$$M_y = \bar{q} \cdot S \cdot c_{ref} \cdot C_{m_{tot}} - x_{cg} \cdot F_z + z_{cg} \cdot F_x.$$

$$M_z = \bar{q} \cdot S \cdot b_{ref} \cdot C_{n_{tot}} + x_{cg} \cdot F_y - y_{cg} \cdot F_x.$$

Appendix C

Appendix for Chapter 3: Part 2

Proof of Theorem 3. 1. *Rewrite Eq. (3.21) in the form of Eq. (3.19), where*

$$\mathbf{H} = \begin{bmatrix} 0 & 0 & \mathbf{P}(\rho) & \mathbf{C}_{acl}^T(\rho) \\ 0 & -\gamma^2 \mathbf{I} & 0 & 0 \\ \mathbf{P}(\rho) & 0 & 0 & 0 \\ \mathbf{C}_{acl}(\rho) & 0 & 0 & -\mathbf{I} \end{bmatrix}, \mathbf{\Psi} = \begin{bmatrix} -\mathbf{A}_{acl}^T(\rho) \\ -\mathbf{G}_a^T(\rho) \\ \mathbf{I} \\ 0 \end{bmatrix}^T,$$

$$\mathbf{\Phi} = \begin{bmatrix} \mathbf{I} & 0 & 0 & 0 \\ 0 & 0 & \mathbf{I} & 0 \end{bmatrix}, \quad \mathbf{M} = \begin{bmatrix} \mathbf{V}(\rho) \\ \mathbf{V}(\rho) \end{bmatrix}.$$

Note that $\mathbf{N}_{\Psi}^T = \begin{bmatrix} \mathbf{I} & 0 & \mathbf{A}_{acl}^T(\rho) & 0 \\ 0 & \mathbf{I} & \mathbf{G}_a^T(\rho) & 0 \\ 0 & 0 & 0 & \mathbf{I} \end{bmatrix}, \mathbf{N}_{\Phi}^T = \begin{bmatrix} 0 & \mathbf{I} & 0 & 0 \\ 0 & 0 & 0 & \mathbf{I} \end{bmatrix}.$

Then, by using Lemma 1(Projection Lemma), Eq. (3.19) is solvable for if and only if Eq. (3.21) holds, the proof is completed.

Proof of Theorem 3. 2. *Based on Eq. (3.21) from Theorem 3.1, Let* $\mathbf{S}(\rho) = \mathbf{V}^{-1}(\rho)$, *pre- and post-multiplying Eq. (3.21) by* $diag\{\mathbf{S}(\rho), \mathbf{I}, \mathbf{S}(\rho), \mathbf{I}\}$ *and* $diag\{\mathbf{S}^T(\rho), \mathbf{I}, \mathbf{S}^T(\rho), \mathbf{I}\}$, *Eq. (3.21) can be rewritten as:*

$$\mathbf{\Pi}(\rho) = \begin{bmatrix} \begin{array}{c} -\mathbf{S}(\rho)\mathbf{A}_{acl}^T(\rho) \\ -\mathbf{A}_{acl}(\rho)\mathbf{S}^T(\rho) \end{array} & -\mathbf{G}_a(\rho) & \begin{array}{c} \mathbf{X}(\rho)+\mathbf{S}(\rho) \\ -\mathbf{S}(\rho)\mathbf{A}_{acl}^T(\rho) \end{array} & \mathbf{S}(\rho)\mathbf{C}_{acl}^T(\rho) \\ * & -\gamma^2\mathbf{I} & -\mathbf{G}_a^T(\rho) & 0 \\ * & * & \mathbf{S}(\rho)+\mathbf{S}^T(\rho) & 0 \\ * & * & * & -\mathbf{I} \end{bmatrix} < 0,$$

(C.1)

Define $\mathbf{L}(\rho) = \mathbf{K}(\rho)\mathbf{S}^T(\rho)$ *where*

$$-\mathbf{S}(\rho)\mathbf{A}_{acl}^T(\rho) - \mathbf{A}_{acl}(\rho)\mathbf{S}^T(\rho) =$$
$$-\left(\mathbf{S}(\rho)\mathbf{A}_a^T(\rho) + \mathbf{L}^T(\rho)\mathbf{B}_a^T(\rho) + \mathbf{A}_a(\rho)\mathbf{S}^T(\rho) + \mathbf{B}_a(\rho)\mathbf{L}(\rho)\right),$$

$$\mathbf{S}(\rho)\mathbf{C}_{acl}^T(\rho) = \mathbf{S}(\rho)\mathbf{C}_z^T(\rho) + \mathbf{L}^T(\rho)\mathbf{D}_z^T(\rho).$$

Eq. (C.1) is equivalent to Eq. (3.21), and $\mathbf{S}(\rho) = \sum_{i=1}^{N} \lambda_i \mathbf{S}_i$, $\mathbf{L}(\rho) = \sum_{i=1}^{N} \lambda_i \mathbf{L}_i$, *and the augmented system matrices described as Eq. (3.13). By expanding every terms of Eq. (C.1), the sufficient condition of Eq. (C.1) holding is*

$$\mathbf{\Pi}\left(\rho\right) = \sum_{i=1}^{N}\sum_{j=1}^{N}\lambda_i\lambda_j\mathbf{\Pi}_{ij} < 0, \tag{C.2}$$

where $\mathbf{\Pi}_{ij}$ is represented as Eq. (3.24).

From Eq. (3.23), we have

$$\mathbf{\Pi}_{ii} < 0, i = 1, \ldots, N, \tag{C.3}$$

$$\mathbf{\Pi}_{ij} + \mathbf{\Pi}_{ji} < 0, 1 \leq i < j \leq N. \tag{C.4}$$

Considering $\sum\limits_{i=1}^{N}\lambda_i = 1$, $\lambda_i \geq 0$, then we can obtain:

$$\mathbf{\Pi}\left(\rho\right) = \sum_{i=1}^{N}\lambda_i^2\mathbf{\Pi}_{ii} + \sum_{i=1}^{N-1}\sum_{j=i+1}^{N}\lambda_i\lambda_j\mathbf{\Pi}_{ij} < 0. \tag{C.5}$$

As can be seen, Eq. (C.5) is equivalent to Eq. (C.2). When Eq. (3.23) holds, then Eq. (C.2) will be satisfied. It is derived that Eq. (C.2) holds and the closed-loop system is stable over all parameter variations ρ. The scheduling parameter λ_i is the function of the parameter ρ, and it could be selected as the method proposed in Eq. (C.1).

Furthermore, the satisfied controller via state feedback is:

$$\mathbf{K}\left(\rho\right) = \left(\sum_{i=1}^{N}\lambda_i\mathbf{L}_i\right)\left(\sum_{i=1}^{N}\lambda_i\mathbf{S}_i^T\right)^{-1}. \tag{C.6}$$

Then, the proof is completed.

Proof of Theorem 3. 3. *By virtue of the structure of $\mathbf{S}\left(\rho\right), \mathbf{L}\left(\rho\right), \mathbf{C}_a\mathbf{T} = \begin{bmatrix}\mathbf{I} & \mathbf{0}\end{bmatrix}$ and Eq. (3.28), we can obtain*

$$\mathbf{L}\left(\rho\right) = \begin{bmatrix}\mathbf{K}_{SOF}\left(\rho\right)\mathbf{S}_{11}\left(\rho\right) & \mathbf{0}\end{bmatrix} = \begin{bmatrix}\mathbf{K}_{SOF}\left(\rho\right) & \mathbf{0}\end{bmatrix}\begin{bmatrix}\mathbf{S}_{11}\left(\rho\right) & \mathbf{0} \\ \mathbf{S}_{21}\left(\rho\right) & \mathbf{S}_{22}\left(\rho\right)\end{bmatrix}$$

$$= \mathbf{K}_{SOF}\left(\rho\right)\begin{bmatrix}\mathbf{I} & \mathbf{0}\end{bmatrix}\mathbf{T}^{-1}\mathbf{T}\mathbf{S}\left(\rho\right) = \mathbf{K}_{SOF}\left(\rho\right)\mathbf{C}_a\mathbf{T}\mathbf{S}\left(\rho\right).$$

From Eqs. (3.26) and (3.27), $\mathbf{\Pi}\left(\rho\right) = \sum\limits_{i=1}^{N}\sum\limits_{j=1}^{N}\lambda_i\lambda_j\mathbf{\Pi}_{ij} < 0$ (for simplicity (ρ) is omitted in the following inequality)

$$
\boldsymbol{\Pi}(\rho) = \begin{bmatrix} \begin{matrix} -(\mathbf{A}_a\mathbf{TS}+\mathbf{B}_a\mathbf{L}) \\ -(\mathbf{A}_a\mathbf{TS}+\mathbf{B}_a\mathbf{L})^T \end{matrix} & -\mathbf{G}_a & \mathbf{Y}+\mathbf{TS}-(\mathbf{A}_a\mathbf{TS}+\mathbf{B}_a\mathbf{L})^T & (\mathbf{C}_z\mathbf{TS}+\mathbf{D}_z\mathbf{L})^T \\ * & -\gamma^2\mathbf{I} & -\mathbf{G}_a^T & 0 \\ * & * & \mathbf{TS}+(\mathbf{TS})^T & 0 \\ * & * & * & -\mathbf{I} \end{bmatrix} < 0.
$$

$$\tag{C.7}$$

Substituting $\mathbf{L}(\rho)$ *for* $\mathbf{K}_{SOF}(\rho)\,\mathbf{C}_a\mathbf{TS}(\rho)$ *in Eq. (C.7), then* $\boldsymbol{\Pi}(\rho)$ *can be described as follows:*

$$
\boldsymbol{\Pi}(\rho) = \begin{bmatrix} \begin{matrix} -(\mathbf{A}_a\mathbf{TS}+\mathbf{B}_a\mathbf{K}_{SOF}\mathbf{CTS}) \\ -(\mathbf{A}_a\mathbf{TS}+\mathbf{B}_a\mathbf{K}_{SOF}\mathbf{CTS})^T \end{matrix} & -\mathbf{G}_a & \begin{matrix} \mathbf{Y}+\mathbf{TS}- \\ \begin{pmatrix} \mathbf{A}_a\mathbf{TS}+ \\ \mathbf{B}_a\mathbf{K}_{SOF}\mathbf{CTS} \end{pmatrix}^T \end{matrix} & \begin{pmatrix} \mathbf{C}_z\mathbf{TS}+ \\ \mathbf{D}_z\mathbf{K}_{SOF}\mathbf{CTS} \end{pmatrix}^T \\ * & -\gamma^2\mathbf{I} & -\mathbf{G}_a^T & 0 \\ * & * & \mathbf{TS}+(\mathbf{TS})^T & 0 \\ * & * & * & -\mathbf{I} \end{bmatrix} < 0.
$$

$$\tag{C.8}$$

Let $\mathbf{W}(\rho) = (\mathbf{TS}(\rho))^{-1}$ *and pre- and post-multiplying Eq. (C.7) by* $\mathrm{diag}\{\mathbf{W}^T,\mathbf{I},\mathbf{W}^T,\mathbf{I}\}$ *and its transpose, it follows that*

$$
\begin{bmatrix} \begin{matrix} -\mathbf{W}^T(\mathbf{A}_a+\mathbf{B}_a\mathbf{K}_{SOF}\mathbf{C}_a) \\ -(\mathbf{A}_a+\mathbf{B}_a\mathbf{K}_{SOF}\mathbf{C}_a)^T\mathbf{W} \end{matrix} & -\mathbf{W}^T\mathbf{G}_a & \begin{matrix} \mathbf{W}^T\mathbf{Y}\mathbf{W}+\mathbf{W}^T- \\ (\mathbf{A}_a+\mathbf{B}_a\mathbf{K}_{SOF}\mathbf{C}_a)^T\mathbf{W} \end{matrix} & \begin{pmatrix} \mathbf{C}_z+ \\ \mathbf{D}_z\mathbf{K}_{SOF}\mathbf{C}_a \end{pmatrix}^T \\ * & -\gamma^2\mathbf{I} & -\mathbf{G}_a^T\mathbf{W} & 0 \\ * & * & \mathbf{W}+\mathbf{W}^T & 0 \\ * & * & * & -\mathbf{I} \end{bmatrix} < 0.
$$

$$\tag{C.9}$$

Let $\mathbf{V}=\mathbf{W}^T$, *then the following inequality is obtained.*

$$
\begin{bmatrix} \begin{matrix} -\mathbf{V}(\mathbf{A}_a+\mathbf{B}_a\mathbf{K}_{SOF}\mathbf{C}_a) \\ -(\mathbf{A}_a+\mathbf{B}_a\mathbf{K}_{SOF}\mathbf{C}_a)^T\mathbf{V}^T \end{matrix} & -\mathbf{V}\mathbf{G}_a & \begin{matrix} \mathbf{P}+\mathbf{V}- \\ \begin{pmatrix} \mathbf{A}_a+ \\ \mathbf{B}_a\mathbf{K}_{SOF}\mathbf{C}_a \end{pmatrix}^T\mathbf{V}^T \end{matrix} & \begin{pmatrix} \mathbf{C}_z+ \\ \mathbf{D}_z\mathbf{K}_{SOF}\mathbf{C}_a \end{pmatrix}^T \\ * & -\gamma^2\mathbf{I} & -\mathbf{G}_a^T\mathbf{V}^T & 0 \\ * & * & \mathbf{V}+\mathbf{V}^T & 0 \\ * & * & * & -\mathbf{I} \end{bmatrix} < 0,
$$

$$\tag{C.10}$$

where $\mathbf{P}=\mathbf{V}\mathbf{X}\mathbf{V}^T$ *is a symmetric matrix. In addition,*

$$
\begin{bmatrix} -\mathbf{V}\mathbf{A}_{acl}-\mathbf{A}_{acl}^T\mathbf{V}^T & -\mathbf{V}\mathbf{G}_a & \mathbf{P}+\mathbf{V}-\mathbf{A}_{acl}^T\mathbf{V}^T & \mathbf{C}_{acl}^T \\ * & -\gamma^2\mathbf{I} & -\mathbf{G}_a^T\mathbf{V}^T & 0 \\ * & * & \mathbf{V}+\mathbf{V}^T & 0 \\ * & * & * & -\mathbf{I} \end{bmatrix} < 0.
$$

Then it is obtained that Eq. (C.7) is equivalent to Eq. (3.21) in Theorem 3.1. From Theorem 3.1, it follows that the closed-loop system is stable.

The sufficient condition of Eq. (C.7) holding is

$$\mathbf{\Pi}(\rho) = \sum_{i=1}^{N} \sum_{j=1}^{N} \lambda_i \lambda_j \mathbf{\Pi}_{ij} < 0, \tag{C.11}$$

where Π_{ij} is represented as Eq. (3.27).

From Eq. (3.26), we have

$$\mathbf{\Pi}_{ii} < 0, i = 1, \ldots, N. \tag{C.12}$$

$$\mathbf{\Pi}_{ij} + \mathbf{\Pi}_{ji} < 0, 1 \leq i < j \leq N. \tag{C.13}$$

Considering $\sum_{i=1}^{N} \lambda_i = 1$, $\lambda_i \geq 0$, then

$$\mathbf{\Pi}_{ij} + \mathbf{\Pi}_{ji} < 0, 1 \leq i < j \leq N. \tag{C.14}$$

Then it is obtained that if Eq. (3.26) holds, then Eq. (C.11) will hold. It is derived that Eq. (3.26) holds and the closed-loop system is stable over all parameter variationsρ. Furthermore, from Eq. (3.27), we can deduce that the matrices TS_i are positive-definite which implies that the matrices $\mathbf{S}_i$ and $\mathbf{S}_{11i}$ are invertible. Then, $\mathbf{L}_1(\rho) = \mathbf{K}_{SOF}(\rho)\mathbf{S}_{11}(\rho)$ admits the solution of Eq. (3.28). The scheduling parameter λ_i is the function of the parameter ρ, and it could be selected as the method proposed in Eq. (C.1). Thus, the proof is completed.

Appendix D

Appendix for Chapter 3: Part 3

1) Normal case.

$$
\mathbf{A}_1 = \begin{bmatrix}
-1.0649 & 0.0034 & -0.0000 & 0.9728 & 0.0000 \\
0.0000 & -0.2492 & 0.0656 & -0.0000 & -0.9879 \\
0.0000 & -22.5462 & -2.0457 & -0.0000 & 0.5432 \\
8.1633 & -0.0057 & -0.0000 & -1.0478 & 0.0000 \\
0.0000 & 1.7970 & -0.1096 & 0.0000 & -0.4357
\end{bmatrix},
$$

$$
\mathbf{B}_1 = \begin{bmatrix}
-0.0062 & -0.0062 & -0.0709 & -0.1172 & -0.1172 & -0.0709 & 0.0003 \\
-0.0072 & 0.0072 & 0.0039 & 0.0188 & -0.0188 & -0.0039 & 0.0627 \\
1.2456 & -1.2456 & -10.6058 & -9.2345 & 9.2345 & 10.6058 & 5.3223 \\
2.7172 & 2.7172 & -2.4724 & -4.0101 & -4.0101 & -2.4724 & 0.0108 \\
-0.7497 & 0.7497 & -0.4923 & -1.1415 & 1.1415 & 0.4923 & -3.7367
\end{bmatrix},
$$

$$
\mathbf{G}_1 = \begin{bmatrix} -0.0072 & 0.0000 & 0.0000 & 0.0551 & -0.0000 \end{bmatrix}^T,
$$

$$
\mathbf{C}_1 = \begin{bmatrix}
1 & 0 & 0 & 0 & 0 \\
0 & 1 & 0 & 0 & 0 \\
0 & 0 & 1 & 0 & 0
\end{bmatrix}.
$$

2) Both the inner elevons loss 100%.

The trimmed values of nominal aircraft are:

$M_a = 0.45$, $h = 3000\,\mathrm{m}$, $V_t = 147.86\,\mathrm{m/s}$, $\alpha = 3.819\,73°$, $\beta = 0$, $T = 0.0752$, $\delta_{rc} = \delta_{lc} = 0.036\,80°$, $\delta_{roe} = \delta_{loe} = 0.015\,92°$, $\delta_r = 0$.

$$
\mathbf{A}_2 = \begin{bmatrix}
-1.0366 & 0.0034 & -0.0000 & 0.9735 & 0.0000 \\
0.0000 & -0.2426 & 0.0670 & -0.0000 & -0.9880 \\
0.0000 & -22.2108 & -1.9911 & -0.0000 & 0.5343 \\
7.9391 & -0.0056 & -0.0000 & -1.0208 & 0.0000 \\
0.0000 & 1.7476 & -0.1077 & 0.0000 & -0.4240
\end{bmatrix},
$$

$$
\mathbf{B}_2 = \begin{bmatrix}
-0.0059 & -0.0059 & -0.0691 & 0.0000 & 0.0000 & -0.0691 & 0.0003 \\
-0.0070 & 0.0070 & 0.0038 & 0.0000 & 0.0000 & -0.0038 & 0.0610 \\
1.2258 & -1.2258 & -10.3167 & 0.0000 & 0.0000 & 10.3167 & 5.1812 \\
2.6460 & 2.6460 & -2.4082 & 0.0000 & 0.0000 & -2.4082 & 0.0105 \\
-0.7324 & 0.7324 & -0.4788 & 0.0000 & 0.0000 & 0.4788 & -3.6368
\end{bmatrix},
$$

$$
\mathbf{G}_2 = \begin{bmatrix} -0.0070 & 0.0000 & 0.0000 & 0.0536 & -0.0000 \end{bmatrix}^T,
$$

179

$$
\mathbf{C_2} = \begin{bmatrix} 1 & 0 & 0 & 0 & 0 \\ 0 & 1 & 0 & 0 & 0 \\ 0 & 0 & 1 & 0 & 0 \end{bmatrix}.
$$

Controller parameters
1) State Feedback Controllers.
The FTC gains:

$$
\mathbf{K}_{S1} = \begin{bmatrix}
10.9466 & 5.0295 & 5.9736 & -3.4314 & -1.6062 & -2.5336 & -0.9582 & 0.5120 \\
10.9588 & -5.0533 & -5.7763 & -3.4357 & 1.6060 & 2.4507 & -0.9591 & -0.5140 \\
-18.5883 & 9.0445 & -88.4660 & 6.5169 & -4.1587 & 36.9424 & 1.0045 & 0.8349 \\
-9.0369 & 0.1786 & 5.6492 & 3.2507 & -0.0376 & -2.3409 & 0.4436 & 0.0611 \\
-8.9845 & -0.1400 & -5.9003 & 3.2323 & 0.0259 & 2.4459 & 0.4406 & -0.0584 \\
-18.6151 & -9.0241 & 88.2960 & 6.5258 & 4.1607 & -36.8710 & 1.0068 & -0.8326 \\
0.0508 & 37.4053 & 26.3773 & -0.0176 & -13.0368 & -11.2000 & -0.0022 & 2.8070
\end{bmatrix},
$$

$$
\mathbf{K}_{S2} = \begin{bmatrix}
13.3230 & 4.9781 & 5.8173 & -4.1728 & -1.5791 & -2.4690 & -1.2658 & 0.5206 \\
13.3337 & -5.0241 & -5.6491 & -4.1763 & 1.5849 & 2.3987 & -1.2671 & -0.5249 \\
-19.7810 & 9.1520 & -78.0220 & 6.8475 & -4.1352 & 32.5975 & 1.2374 & 0.9168 \\
-8.9746 & 0.1589 & 5.6004 & 3.2450 & -0.0295 & -2.3209 & 0.4813 & 0.0588 \\
-8.9126 & -0.1076 & -5.8550 & 3.2233 & 0.0130 & 2.4272 & 0.4773 & -0.0549 \\
-19.7779 & -9.1176 & 77.8872 & 6.8462 & 4.1338 & -32.5411 & 1.2372 & -0.9129 \\
0.0627 & 36.8939 & 26.1832 & -0.0219 & -12.7885 & -11.1231 & -0.0024 & 2.8466
\end{bmatrix}.
$$

The reliable controller gain:

$$
\mathbf{K}_{rel} = \begin{bmatrix}
12.3083 & 4.8827 & 6.1672 & -3.7755 & -1.5403 & -2.6159 & -1.2195 & 0.5190 \\
12.3234 & -4.9261 & -6.1325 & -3.7803 & 1.5446 & 2.6016 & -1.2212 & -0.5242 \\
-19.1874 & 8.5590 & -78.3197 & 6.5822 & -3.9160 & 32.7020 & 1.2230 & 0.8696 \\
-1.2110 & 0.1222 & 2.7767 & 0.4069 & -0.0290 & -1.1648 & 0.0756 & 0.0078 \\
-1.2043 & -0.1424 & -2.8526 & 0.4047 & 0.0362 & 1.1968 & 0.0752 & -0.0101 \\
-19.1935 & -8.5129 & 78.2900 & 6.5846 & 3.9103 & -32.6897 & 1.2229 & -0.8649 \\
0.0563 & 36.4327 & 29.0835 & -0.0211 & -12.5992 & -12.3419 & -0.0011 & 2.8422
\end{bmatrix}.
$$

The robust controller gain:

$$
\mathbf{K}_{rob} = \begin{bmatrix}
7.5442 & 5.5339 & -1.5210 & -2.3547 & -1.7912 & 0.6081 & -0.6469 & 0.5185 \\
7.5495 & -5.5593 & 1.6202 & -2.3563 & 1.7951 & -0.6495 & -0.6475 & -0.5205 \\
-15.2466 & 4.1893 & -80.3349 & 5.4953 & -2.0369 & 33.5069 & 0.6738 & 0.3856 \\
-24.9738 & 11.5568 & -77.9905 & 9.0108 & -4.6088 & 32.5061 & 1.0956 & 0.8769 \\
-24.9732 & -11.5251 & 77.8612 & 9.0108 & 4.6066 & -32.4522 & 1.0953 & -0.8737 \\
-15.2502 & -4.1697 & 80.2550 & 5.4967 & 2.0355 & -33.4736 & 0.6740 & -0.3837 \\
0.0539 & 39.4809 & -6.1890 & -0.0199 & -13.7759 & 2.4525 & -0.0015 & 2.7848
\end{bmatrix}.
$$

2) Static Output Feedback Controllers
The FTC gains:

$$
\mathbf{K}_{SOF1} = \begin{bmatrix}
-1.6682 & -1.2416 & 0.2839 & 1.1943 & 3.1957 & -0.1174 \\
-1.6648 & 1.2399 & -0.2867 & 1.1897 & -3.1931 & 0.1188 \\
-1.1999 & -0.6467 & -3.3716 & 1.6546 & 0.2312 & 1.7493 \\
-0.8026 & -0.2179 & -1.1903 & 1.1610 & 0.3455 & 0.6366 \\
-0.8025 & 0.2181 & 1.1911 & 1.1605 & -0.3459 & -0.6371 \\
-1.1982 & 0.6485 & 3.3736 & 1.6526 & -0.2342 & -1.7506 \\
0.0010 & 0.8890 & 1.5015 & -0.0014 & -1.2899 & -0.8354
\end{bmatrix}.
$$

$$\mathbf{K}_{\text{SOF2}} = \begin{bmatrix} -1.2120 & -1.3696 & 0.2806 & 0.5013 & 3.6383 & -0.1046 \\ -1.2065 & 1.3677 & -0.2832 & 0.4943 & -3.6350 & 0.1060 \\ -1.9304 & -0.9232 & -4.6317 & 2.7485 & 0.6145 & 2.4297 \\ -0.9430 & -0.2493 & -1.2582 & 1.3131 & 0.4056 & 0.6702 \\ -0.9427 & 0.2494 & 1.2591 & 1.3125 & -0.4059 & -0.6707 \\ -1.9278 & 0.9249 & 4.6334 & 2.7449 & -0.6174 & -2.4308 \\ 0.0015 & 0.8784 & 1.1035 & -0.0019 & -1.3604 & -0.6292 \end{bmatrix}.$$

The reliable controller gain:

$$\mathbf{K}_{rel\_SOF} = \begin{bmatrix} -1.1789 & -1.3554 & 0.3795 & 0.4637 & 3.6033 & -0.1633 \\ -1.1730 & 1.3533 & -0.3838 & 0.4572 & -3.5995 & 0.1657 \\ -1.9088 & -1.3334 & -6.7009 & 2.7188 & 1.2945 & 3.5417 \\ 0.0007 & -0.0048 & 0.0023 & -0.0057 & 0.0156 & -0.0004 \\ 0.0007 & 0.0051 & -0.0030 & -0.0058 & -0.0153 & 0.0009 \\ -1.9071 & 1.3353 & 6.7040 & 2.7165 & -1.2979 & -3.5435 \\ 0.0015 & 0.9811 & 1.6101 & -0.0017 & -1.5260 & -0.8991 \end{bmatrix}.$$

The robust controller gain:

$$\mathbf{K}_{rob\_SOF} = \begin{bmatrix} -2.1106 & -1.2668 & 0.5842 & 1.7076 & 3.3491 & -0.2722 \\ -2.1075 & 1.2657 & -0.5873 & 1.7040 & -3.3475 & 0.2737 \\ -0.7927 & -0.9022 & -3.8848 & 1.0967 & 1.0655 & 2.0622 \\ -1.3634 & -0.3857 & -3.2298 & 1.8548 & 0.0672 & 1.6988 \\ -1.3618 & 0.3872 & 3.2336 & 1.8528 & -0.0693 & -1.7009 \\ -0.7912 & 0.9032 & 3.8871 & 1.0952 & -1.0669 & -2.0636 \\ 0.0014 & 1.0694 & 2.7612 & -0.0016 & -1.4019 & -1.5042 \end{bmatrix}.$$

Appendix E

Appendix for Chapter 4

E.1 Experimental Parameters

Related physical parameters of the quadrotor UAV, and some selected gains of our proposed controllers and observers are presented as follows.

E.1.1 Physical Parameters

$m = 1.121\,\text{kg}$, $I_{xx} = 0.01\,\text{kg} \cdot \text{m}^2$, $I_{yy} = 0.0082\,\text{kg} \cdot \text{m}^2$, $I_{zz} = 0.0148\,\text{kg} \cdot \text{m}^2$, $d_\theta = 0.0879\,\text{m}$, $d_\phi = 0.1068\,\text{m}$, $c_{\tau f} = 0.00963$.

E.1.2 Gains

K_γ, K_ν in Eq. (4.8) are selected as

$$K_\gamma = \begin{bmatrix} 12 & 0 & 0 \\ 0 & 12 & 0 \\ 0 & 0 & 35 \end{bmatrix}, K_\nu = \begin{bmatrix} 8 & 0 & 0 \\ 0 & 8 & 0 \\ 0 & 0 & 18 \end{bmatrix}.$$

K_η, K_ω in Eq. (4.15) are selected as

$$K_\eta = \begin{bmatrix} 2.16 & 0 & 0 \\ 0 & 1.92 & 0 \\ 0 & 0 & 0.59 \end{bmatrix}, K_\omega = \begin{bmatrix} 0.20 & 0 & 0 \\ 0 & 0.12 & 0 \\ 0 & 0 & 0.12 \end{bmatrix}.$$

K_p and $l(\gamma, \nu)$ in Eqs. (4.12) and (4.9) are chosen as

$$K_p = \begin{bmatrix} 50 & 0 & 0 & 833 & 0 & 0 & 78 & 0 & 0 \\ 0 & 50 & 0 & 0 & 833 & 0 & 0 & 78 & 0 \\ 0 & 0 & 50 & 0 & 0 & 833 & 0 & 0 & 78 \end{bmatrix},$$

$$l(\gamma, \nu) = \begin{bmatrix} 0 & 0 & 0 & 0.1 & 0 & 0 \\ 0 & 0 & 0 & 0.1 & 0 & 0 \\ 0 & 0 & 0 & 0 & 0.08 & 0 \\ 0 & 0 & 0 & 0 & 0.08 & 0 \\ 0 & 0 & 0 & 0 & 0 & 0.1 \\ 0 & 0 & 0 & 0 & 0 & 0.1 \end{bmatrix}.$$

K_a in Eq. (4.17) is chosen as

$$K_a = \begin{bmatrix} 50 & 0 & 0 & 833 & 0 & 0 & 3906 & 0 & 0 \\ 0 & 50 & 0 & 0 & 833 & 0 & 0 & 3906 & 0 \\ 0 & 0 & 50 & 0 & 0 & 833 & 0 & 0 & 3906 \end{bmatrix}.$$

Bibliography

[1] M. Blanke, R. Izadi Zamanabadi, R. Bogh, Z.P Lunau. Fault-tolerant control systems – A holistic view. *Control Eng. Pract.*, 5(5):693–702, 1997.

[2] J. Jiang. Fault-tolerant control systems – An introductory overview. *Automatica SINCA*, 21(1):161–174, 2005.

[3] R.J. Patton. Fault-tolerant control systems: The 1997 situation. In *Proceedings of the 3rd IFAC Symposium on Fault Detection, Supervision and Safety for Technical Processes*, pages 1033–1055. Kingston upon Hull, UK, 1997.

[4] Y.M. Zhang, J. Jiang. Bibliographical review on reconfigurable fault-tolerant control. *Annu. Rev. Control*, 32(2):229–252, 2008.

[5] H. Noura, D. Sauter, F. Hamelin, D. Theilliol. Fault-tolerant control in dynamic systems: Application to a winding machine. *IEEE Control Syst. Mag.*, 20(1):117–124, 2000.

[6] J.S. Brinker, K.E. Wise. Flight testing of reconfigurable control law on the x-36 tailless aircraft. *AIAA J. Guid. Control Dyn.*, 24(5):903–909, 2001.

[7] L. Ye, S. Wang, F. Bing, O.P. Malik, Y. Zeng. Control/Maintenance strategy fault tolerant mode and reliability analysis for hydro power stations. *IEEE Trans. Power Syst.*, 16(3):340–345, 2001.

[8] Y.M. Zhang, J. Jiang. Integrated design of reconfigurable fault tolerant control systems. *AIAA J. Guid. Control Dyn.*, 24(1):133–136, 2001.

[9] M. Maki, J. Jiang, K. Hagino. A stability guaranteed active fault-tolerant control system against actuator failures. *Int. J. Robust Nonlinear Control*, 14(12):1061–1077, 2004.

[10] M.M. Kale, A.J. Chipperfield. Stabilized MPC formulations for robust reconfigurable flight control. *Control Eng. Pract.*, 13(6):771–788, 2005.

[11] J. Jiang, Y.M. Zhang. Accepting performance degradation in fault-tolerant control system design. *IEEE Trans. Control Syst. Technol.*, 14(2):284–292, 2006.

[12] H. Alwi, C. Edwards. Fault tolerant control using sliding modes with on-line control allocation. *Automatica*, 44(7):1859–1866, 2008.

[13] H. Ltaief, E. Gabriel, M. Garbey. Fault tolerant algorithms for heat transfer problems. *J. Parallel Distrib. Comput.*, 68:663–677, 2008.

[14] H. Alwi, C. Edwards, C.P. Tan. Sliding mode estimation schemes for incipient sensor faults. *Automatica*, 45(7):1679–1685, 2009.

[15] F.N. Pirmoradi, F. Sassani, C.W. de Silva. Fault detection and diagnosis in a spacecraft attitude determination system. *ACTA Astronautica*, 65(5–6):710–729, 2009.

[16] S. Varma, K.D. Kumar. Fault tolerant satellite attitude control using solar radiation pressure based on nonlinear adaptive sliding mode. *ACTA Astronautica*, 66(3–4):710–729, 2010.

[17] A. Zolghadri. On flight operational issues management: Past, present and future. *Annu. Rev. Control*, 45:41–51, 2018.

[18] J.C. Geromel, J. Bernussou, M.C. de Oliveira. H_2 norm optimization with constrained dynamic output feedback controllers: Decentralized and reliable control. *IEEE Trans. Autom. Control*, 38(11):1985–1990, 2002.

[19] C.S. Hsieh. Performance gain margins of the two-stage LQ reliable control. *Automatica*, 44(7):1449–1454, 1999.

[20] J. Jiang, Q. Zhao. Design of reliable control systems possessing actuator redundancies. *AIAA J. Guid. Control Dyn.*, 23(4):709–718, 2000.

[21] F. Liao, J.L. Wang, G.H. Yang. Reliable robust flight tracking control: An LMI approach. *IEEE Trans. Control Syst. Technol.*, 10(1):76–89, 2002.

[22] C.J. Seo, B.K. Kim. Robust and reliable H_∞ control for linear systems with parameter uncertainty and actuator failure. *Automatica*, 32(3):465–467, 1996.

[23] R.J. Veillette. Reliable linear-quadratic state-feedback control. *Automatica*, 31(1):137–143, 1995.

[24] R.J. Veillette, S.V. Medanic, W.R. Perkins. Design of reliable control systems. *IEEE Trans. Autom. Control*, 37(3):290–304, 1992.

[25] G.H. Yang, J.L. Wang, Y.C, Soh. Reliable H_∞ controller design for linear systems. *Automatica*, 37(5):717–725, 2001.

[26] G.H. Yang, D. Ye. Reliable H_∞ control of linear systems with adaptive mechanism. *IEEE Trans. Autom. Control*, 55(1):242–247, 2010.

[27] Q. Zhao, J. Jiang. Reliable state feedback control systems design against actuator failures. *Automatica*, 34(10):1267–1272, 1998.

[28] L. Guo, S. Cao. Anti-disturbance control theory for systems with multiple disturbances: A survey. *ISA Trans.*, 53(4):846–849, 2014.

[29] J. Jiang, Q. Zhao. Comparison of parameter and state estimation based FDI algorithms. In *Proceedings of the 11th IFAC Symposium on System Identification*, pages 655–660. Stockholm, Sweden, 1997.

[30] X. Yu, J. Jiang. Hybrid fault-tolerant flight control system design against partial actuator failures. *IEEE Trans. Control Syst. Technol.*, 20(4):871–886, 2012.

[31] T. Nagashio, T. Kida, Y. Hamada, T. Ohtani. Robust two-degrees-of-freedom attitude controller design and flight test result for engineering test satellite-VIII spacecraft. *IEEE Trans. Control Syst. Technol.*, 22(1):157–168, 2014.

[32] Y. Bai, J.D. Biggs, F.B. Zazzera, N.G. Cui. Adaptive attitude tracking with active uncertainty rejection. *AIAA J. Guid. Control Dyn.*, 41(2):546–554, 2017.

[33] J.Q. Han. From PID to active disturbance rejection control. *IEEE Trans. Ind. Electron.*, 56(3):900–906, 2009.

[34] Y. Zhang, Z. Chen, X. Zhang, Q. Sun, M. Sun. A novel control scheme for quadrotor UAV based upon active disturbance rejection control. *Aerosp. Sci. Technol.*, 79:601–609, 2018.

[35] L. Guo, W. Chen. Disturbance attenuation and rejection for systems with nonlinearity via DOBC approach. *Int. J. Robust Nonlinear Control*, 15(3):109–125, 2005.

[36] W.H. Chen, J. Yang, L. Guo, S.H. Li. Disturbance-observer-based control and related methods—An overview. *IEEE Trans. Ind. Electron.*, 63(2):1083–1095, 2016.

[37] H. Lu, C. Liu, L. Guo, W. Chen. Flight control design for small-scale helicopter using disturbance observer based backstepping. *AIAA J. Guid. Control Dyn.*, 38(11):2235–2240, 2015.

[38] L. Guo, S. Cao. *Anti-Disturbance Control for Systems with Multiple Disturbances*. CRC Press, Florida, 2013.

[39] J. Jiang, Q. Zhao. Fault tolerant control system synthesis using imprecise fault identification and reconfiguration control. In *Proceedings of IEEE International Symposium on Intelligent Control*, pages 169–174. Gaithersburg, USA, 1998.

[40] J. Jiang, Q. Zhao. Reconfigurable control based on imprecise fault identification. In *Proceedings of American Control Conference (ACC)*, pages 114–118. San Diego, USA, 1999.

[41] M. Blanke, M. Kinnaert, J. Lunze, M. Staroswiecki. *Diagnosis and Fault-Tolerant Control*. Springer-Verlag, Berlin, 2006.

[42] J. Jiang, X. Yu. Fault-tolerant control systems: A comparative study between active and passive approaches. *Annu. Rev. Control*, 36(1):60–72, 2012.

[43] J.R. Sklaroff. Redundancy management technique for space shuttle computers. *IBM J. Res. Dev.*, 20(1):20–28, 1976.

[44] K.F. Lu, Y.Q. Xia. Finite-time fault-tolerant control for rigid spacecraft with actuator saturations. *IET Control Theory Appl.*, 7(11):1529–1539, 2013.

[45] M. Staroswiecki. On reconfigurability with respect to actuator failures. In *Proceedings of the 15th IFAC World Congress*, pages 772–777. Barcelona, Spain, 2002.

[46] M. Guler, S. Clements, L.M. Wills, B.S. Heck, G.J. Vachtsevanos. Transition management for reconfigurable hybrid control systems. *IEEE Control Syst. Mag.*, 23(1):36–49, 2003.

[47] J. Cieslak, D. Efimov, D. Henry. Transient management of a supervisory fault-tolerant control scheme based on dwell-time conditions. *Int. J. Adapt Control Signal Process*, 29(1):123–142, 2015.

[48] D. Bustan, S. Sani, N. Pariz. Adaptive fault-tolerant spacecraft attitude control design with transient response control. *IEEE/ASME Trans. Mechatron.*, 19(4):1404–1411, 2014.

[49] Y.M. Zhang, J. Jiang. Fault tolerant control system design with explicit consideration of performance degradation. *IEEE Trans. Aerosp. Electron. Syst.*, 39(3):838–848, 2003.

[50] M. Benosman, K.Y. Lum. Online references reshaping and control reallocation for nonlinear fault tolerant control. *IEEE Trans. Control Syst. Technol.*, 17(2):366–379, 2009.

[51] X.X. Hu, H.R. Karimi, L.G. Wu, Y. Guo. Model predictive control-based nonlinear fault tolerant control for air-breathing hypersonic vehicles. *IET Control Theory Appl.*, 8(13):1147–1153, 2014.

[52] F. Bateman, H. Noura, M. Ouladsine. Fault diagnosis and fault-tolerant control strategy for the aerosonde UAV. *IEEE Trans. Aerosp. Electron. Syst.*, 47(3):2119–2137, 2011.

[53] J.H. Fan, Y.M. Zhang, Z.Q. Zheng. Robust fault-tolerant control against time-varying actuator faults and saturation. *IET Control Theory Appl.*, 6(14):2198–2208, 2012.

[54] J.H. Fan, Y.M. Zhang, Z.Q. Zheng. Adaptive observer-based integrated fault diagnosis and fault-tolerant control systems against actuator faults and saturation. *J. Dyn. Syst. Meas. Contr.*, 135(4):041008-1–041008-13, 2013.

[55] X. Yu, Y.M. Zhang, Z.X. Liu. Fault-tolerant flight control design with explicit consideration of reconfiguration transients. *AIAA J. Guid. Control Dyn.*, 39(3):556–563, 2016.

[56] X. Yu, Z.X. Liu, Y.M. Zhang. Fault-tolerant flight control design with finite-time adaptation under actuator stuck failures. *IEEE Trans. Control Syst. Technol.*, 25(4):1431–1440, 2017.

[57] E. Balaban, A. Saxena, P. Bansal, K.F. Geobel, S. Curran. Modeling, detection, and disambiguation of sensor faults for aerospace applications. *IEEE Sensors J.*, 9(12):1907–1917, 2009.

[58] M.K. Jeerage. Reliability analysis of fault-tolerant IMU architectures with redundant inertial sensors. *IEEE Aerosp. Electron. Syst. Mag.*, 5(7):23–28, 1990.

[59] S. Osder. Practical view of redundancy management application and theory. *AIAA J. Guid. Control Dyn.*, 22(1):12–21, 1999.

[60] X. Yu, Y. Fu, Y.M Zhang. Aircraft fault accommodation with consideration of actuator control authority and gyro availability. *IEEE Trans. Control Syst. Technol.*, 26(4):1285–1299, 2018.

[61] M. Steinberg. Historical overview of research in reconfigurable flight control. In *Proceedings of the Institution of Mechanical Engineers, Part G: Journal of Aerospace Engineering*, pages 263–275, 2005.

[62] J. Cicslak, D. Henry, A. Zolghadri. Fault tolerant flight control: From theory to piloted flight simulator experiments. *IET Control Theory Appl.*, 4(8):1451–1464, 2010.

[63] H.H. Niemann, J. Stoustrup. An architecture for fault tolerant controllers. *Int. J. Control*, 78(14):1091–1110, 2005.

[64] H.H. Niemann, N.K. Poulsen. A concept for fault tolerant controller. In *Proceedings of Diagnosis of Processes and Systems*, pages 107–114, 2009.

[65] H.H. Niemann, N.K. Poulsen. Fault tolerant control—A residual based set-up. In *Proceedings of the 48th IEEE Conference on Decision and Control (CDC)*, pages 8470–8475. Gdansk, Poland, 2009.

[66] H.H. Niemann. A model-based approach for fault-tolerant control. *Int. J. Appl. Math. Comput. Sci.*, 22(1):67–86, 2012.

[67] Y.M. Zhang, J. Jiang. Active fault-tolerant control system against partial actuator failures. *IEE Proc. Control Theory Appl.*, 149(1):95–104, 2002.

[68] H.Z. Tan, N. Sepehri. Parametric fault diagnosis for electrohydraulic cylinder drive units. *IEEE Trans. Ind. Electron.*, 49(1):96–106, 2002.

[69] D.F. Thompson, J.S. Pruyn, A. Shukla. Feedback design for robust tracking and robust stiffness in flight control actuators using a modified QFT technique. *Int. J. Control*, 72(16):1480–1497, 1999.

[70] N. Niksefat, N. Sepehri. A QFT fault-tolerant control for electrohydraulic positioning systems. *IEEE Trans. Control Syst. Technol.*, 10(4):626–632, 2002.

[71] M. Karpenko, N. Sepehri. A QFT fault-tolerant control for electrohydraulic positioning systems. *Mechatronics*, 19(7):1067–1077, 2009.

[72] M. Karpenko, N. Sepehri. Fault-tolerant control of a servohydraulic positioning system with crossport leakage. *IEEE Trans. Control Syst. Technol.*, 13(1):155–161, 2005.

[73] B.L. Stevens, F.L. Lewis. *Aircraft Control and Simulation*. Wiley, New York, 1992.

[74] S.P. Boyd, E. Ghaoui, L. Feron, V. Balakrishnan. *Linear Matrix Inequalities in System and Control Theory*. SIAM, Philadelphia, PA, 1994.

[75] P. Apkarian, H.D. Tuan, J. Bernussou. Continuous-time analysis and H_2 multi-channel synthesis with enhanced LMI characterizations. *IEEE Trans. Autom. Control*, 46(12):1941–1946, 2001.

[76] L. Forsell, U. Nilsson. *ADMIRE the aero-data model in a research environment Version 4.0, model description*. Stockholm, Sweden, Tech. Rep. FOI-R-1624-SE, 2005.

[77] R.A. Eslinger, P.R. Chandler. Self-repairing flight control system program overview. In *Proceedings of the IEEE NAECON*, pages 504–512, 1988.

[78] U.S. General Aviatio. *Annual Review of General Aviation Accident Data 2005*. Washington, D.C., NTSB/ARG-09/01, 2009.

[79] J.F. Stewart, T.L. Shuck. *Flight-testing of the self-repairing flight control system using the F-15 highly integrated digital electronic control flight research facility*. Ames Research Center, Dryden Flight Research Facility, Edwards, CA, NASA Technical Report, 1990.

[80] F. Liao, J.L. Wang, G.H. Yang. Reliable aircraft tracking control via quadratic parameter dependent Lyapunov functions: State feedback case. In *Proceedings of the CDC*, pages 4474–4479, 2002.

[81] F. Liao, J.L. Wang, G.H. Yang. Reliable H_2 static output feedback tracking control against aircraft wing/control surface impairment. In *2003 IEEE International Workshop on Workload Characterization*, pages 112–117. Petersburg, Russia, 2003.

[82] L. Feng, et al. Reliable H_∞ aircraft flight controller design against faults with state/output feedback. In *Proceedings of the ACC*, pages 2664–2669, 2005.

[83] M. Rodrigues, et al. Fault tolerant control design for polytopic LPV systems. *Int. J. Appl. Math. Comput. Sci.*, 17(1):27–37, 2007.

[84] P. Apkarian, P. Gahinet, G. Becker. Self-scheduled H_∞ control of linear parameter-varying systems: A design example. *Automatica*, 31(9):1251–1261, 1995.

[85] H.H. Rosenbrock. *State Space and Multivariable Theory*. Nelson, London, 1970.

[86] M.R. Napolitano, Y. Song, B. Seanor. On-line parameter estimation for restructurable flight control systems. *Aircraft Design*, 4(1):19–50, 2001.

[87] P. Gahinet, P. Apkarian. Output-feedback-based H_∞ control for vehicle suspension systems with control delay. *Int. J. Robust Nonlinear Control*, 4(4):421–448, 1994.

[88] M.R. Napolitano, Y. Song, B. Seanor. A fault tolerant flight control system for sensor and actuator failures using neural networks. *Aircraft Design*, 3(2):103–128, 2000.

[89] K.M. Zhou, J.C. Doyle. *Essentials of Robust Control*. Prentice-Hall, Upper Saddle River, NJ, 1998.

[90] K. Guo, X. Li, L. Xie. Ultra-wideband and odometry-based cooperative relative localization with application to multi-UAV formation control. *IEEE Trans. Cybern.*, 50(6):2590–2603, 2020.

[91] G.V. Raffo, M.G. Ortega, F.R. Rubio. An integral predictive/nonlinear H_∞ control structure for a quadrotor helicopter. *Automatica*, 46(1):29–39, 2010.

[92] J. Xu, P. Shi, C. Lim, C. Cai, Y. Zou. Reliable tracking control for under-actuated quadrotors with wind disturbances. *IEEE Trans. Syst. Man Cybern.: Syst.*, 49(10):2059–2070, 2019.

[93] Y. Chen, Y. He, M. Zhou. Decentralized PID neural network control for a quadrotor helicopter subjected to wind disturbance. *J. Cent. South Univ.*, 22(1):168–179, 2015.

[94] K. Alexis, G. Nikolakopoulos, A. Tzes. Switching model predictive attitude control for a quadrotor helicopter subject to atmospheric disturbances. *Control Eng. Pract.*, 19(10):1195–1207, 2011.

[95] M. Hehn, R. D'Andrea. A frequency domain iterative learning algorithm for high-performance, periodic quadrocopter maneuvers. *Mechatronics*, 24(8):954–965, 2014.

[96] L. Besnard, Y.B. Shtessel, B. Landrum. Quadrotor vehicle control via sliding mode controller driven by sliding mode disturbance observer. *J. Franklin Inst.*, 349(2):658–684, 2012.

[97] S. Waslander, C. Wang. Wind disturbance estimation and rejection for quadrotor position control. In *AIAA Infotech@Aerospace Conference*, pages 1–14. Seattle, USA, 2009.

[98] W. Dong, G.Y. Gu, X. Zhu, H. Ding. High-performance trajectory tracking control of a quadrotor with disturbance observer. *Sens. Actuators, A.*, 211:67–77, 2014.

[99] X. Lyu, J. Zhou, H. Gu, Z. Li, S. Shen, F. Zhang. Disturbance observer based hovering control of quadrotor tail-sitter VTOL UAVs using H_∞ synthesis. *IEEE Robot. Autom. Lett.*, 3(4):2910–2917, 2018.

[100] W. Chen, D.J. Ballance, P.J. Gawthrop, J. O'Reilly. A nonlinear disturbance observer for robotic manipulators. *IEEE Trans. Ind. Electron.*, 47(4):932–938, 2000.

[101] L. Guo, X. Wen. Hierarchical anti-disturbance adaptive control for nonlinear systems with composite disturbances and applications to missile systems. *Trans. Inst. Meas. Control*, 33(8):942–956, 2011.

[102] Z. Gao. Scaling and bandwidth-parameterization based controller tuning. In *Proceedings of the 2003 American Control Conference*, pages 4989–4996. Minneapolis, USA, 2006.

[103] Y. Yuan, L. Cheng, Z. Wang, C. Sun. Position tracking and attitude control for quadrotors via active disturbance rejection control method. *Science in China Series F: Information Sciences*, 62(1):10201, 2019.

[104] Y. Zhu, L. Guo, J. Qiao, W. Li. An enhanced anti-disturbance attitude control law for flexible spacecrafts subject to multiple disturbances. *Control Eng. Pract.*, 84:274–283, 2019.

[105] Y. Guo, B. Jiang, Y. Zhang. A novel robust attitude control for quadrotor aircraft subject to actuator faults and wind gusts. *IEEE/CAA J. Automatica Sinica*, 5(1):292–300, 2018.

[106] W. Chen. Disturbance observer based control for nonlinear systems. *IEEE/ASME Trans. Mechatron.*, 9(4):706–710, 2004.

[107] W. Chen. Nonlinear disturbance observer-enhanced dynamic inversion control of missiles. *AIAA J. Guid. Control Dyn.*, 26(1):161–166, 2003.

[108] B.N. Pamadi, P.F. Covell, P.V. Tartabini. Aerodynamic characteristics and glide-back performance of Langley glide-back booster. In *Proceedings of the 22nd Applied Aerodynamics Conference and Exhibit*, pages 1–17. Providence, USA, 2004.

[109] C. Michael, B. Peter, K. Michael. Adaptive control for a hypersonic glider using parameter feedback from system identification. In *Proceedings of AIAA Guidance Navigation, and Control Conference*, pages 1–18. Portland, USA, 2011.

[110] S. Banerjee, Z.J. Wang, B. Baur, F. Holzapfel, J.X. Che, C.Y. Cao. L_1 adaptive control augmentation for the longitudinal dynamics of a hypersonic glider. *AIAA J. Guid. Control Dyn.*, 39(2):275–291, 2016.

[111] L. Fiorentini, A. Serrani, M.A. Bolender, D.B. Doman. Nonlinear robust adaptive control of flexible air-breathing hypersonic vehicles. *AIAA J. Guid. Control Dyn.*, 32(2):402–417, 2009.

[112] L.G. Wu, X. Yang, F. Li. Nonfragile output tracking control of hypersonic air-breathing vehicles with an LPV model. *IEEE/ASME Trans. Mechatron.*, 18(4):1280–1288, 2013.

[113] H.J. Xu, M.D. Mirmirani, P.A. Ioannou. Adaptive sliding mode control design for a hypersonic flight vehicle. *AIAA J. Guid. Control Dyn.*, 27(5):829–838, 2004.

[114] J. Yang, Z.H. Zhao, S.H. Li, W.X. Zheng. Composite predictive flight control for airbreathing hypersonic vehicles. *Int. J. Control*, 87(9):1970–1984, 2014.

[115] L. Fiorentini, A. Serrani. Adaptive restricted trajectory tracking for a non-minimum phase hypersonic vehicle model. *Automatica*, 48(7):1248–1261, 2012.

[116] P. Yu, Y. Shtessel. Continuous higher order sliding mode control with adaptation of air breathing hypersonic missile. *Int. J. Adapt Control Signal Process.*, 30(8):1099–1117, 2016.

[117] Q.K. Shen, B. Jiang, V. Cocquempot. Fault-tolerant control for T–S fuzzy systems with application to near-space hypersonic vehicle with actuator faults. *IEEE Trans. Fuzzy Syst.*, 20(4):652–665, 2012.

[118] Q.K. Shen, B. Jiang, V. Cocquempot. Fuzzy logic system-based adaptive fault-tolerant control for near-space vehicle attitude dynamics with actuator faults. *IEEE Trans. Fuzzy Syst.*, 21(2):289–300, 2013.

[119] F. Wu, X. Cai. Switching fault tolerant control of a flexible air-breathing hypersonic vehicle. In *Proceedings of the Institution of Mechanical Engineers, Part I: Journal of Systems and Control Engineering*, pages 24–38, 2013.

[120] J. Zhao, B. Jiang, P. Shi, Z.F. Gao, D.Z. Xu. Fault-tolerant control design for near-space vehicles based on a dynamic terminal sliding mode techniques. In *Proceedings of the Institution of Mechanical Engineers, Part I: Journal of Systems and Control Engineering*, pages 787–794, 2012.

[121] Y.H. Ji, H.L. Zhou, Q. Zong. Adaptive active fault-tolerant control of generic hypersonic flight vehicles. In *Proceedings of the Institution of Mechanical Engineers, Part I: Journal of Systems and Control Engineering*, pages 130–138, 2015.

[122] H. An, J.X. Liu, C.H. Wang, L.G. Wu. Approximate back-stepping fault-tolerant control of the flexible air-breathing hypersonic vehicle. *IEEE/ASME Trans. Mechatron.*, 21(3):1680–1691, 2016.

[123] Z.H. Man, A.P. Paplinski, H.R. Wu. A robust MIMO terminal sliding mode control scheme for rigid robotic manipulator. *IEEE Trans. Autom. Control*, 39(12):2464–2469, 1994.

[124] J. Zheng, H. Wang, Z.H. Man, J. Jin, M.Y. Fu. Robust motion control of a linear motor positioner using fast nonsingular terminal sliding mode. *IEEE/ASME Trans. Mechatron.*, 20(4):1743–1752, 2015.

[125] M. Jin, J. Lee, K. Ahn. Continuous nonsingular terminal sliding-mode control of shape memory alloy actuators using time delay estimation. *IEEE/ASME Trans. Mechatron.*, 20(2):899–909, 2015.

[126] Z.F. Gao, B. Jiang, P. Shi, M.S. Qian, J.X. Lin. Active fault tolerant control design for reusable launch vehicle using adaptive sliding mode technique. *J. Franklin Inst.*, 349(4):1543–1560, 2012.

[127] H.B. Sun, S.H. Li, C.Y. Sun. Adaptive active fault-tolerant control of generic hypersonic flight vehicles. In *Proceedings of the Institution of Mechanical Engineers, Part I: Journal of Systems and Control Engineering*, pages 1344–1355, 2012.

[128] T. Cao, Z.F. Gao, M.S. Qian, J. Zhao. Passive fault tolerant control approach for hypersonic vehicle with actuator loss of effectiveness faults. In *Proceedings of the 28th Chinese Control and Decision Conference (CCDC)*, pages 5951–5956. Yinchuan, China, 2016.

[129] X. Yu, J. Jiang. A survey of fault-tolerant controllers based on safety-related issues. *Annu. Rev. Control*, 39(1):46–57, 2015.

[130] Y. Shtessel, J. Buffington, S. Banda. Multiple timescale flight control using reconfigurable sliding modes. *AIAA J. Guid. Control Dyn.*, 22(6):873–883, 1999.

[131] B.L. Tian, W.R. Fan, R. Su, Q. Zong. Real-time trajectory and attitude coordination control for reusable launch vehicle in reentry phase. *IEEE Trans. Ind. Electron.*, 62(3):1639–1650, 2015.

[132] S. Bhat, D. Bernstein. Finite-time stability of continuous autonomous systems. *SIAM J. Control Optim.*, 38(3):751–766, 2000.

[133] B. Xiao, S. Yin. Velocity-free fault-tolerant and uncertainty attenuation control for a class of nonlinear systems. *IEEE Trans. Ind. Electron.*, 63(7):4400–4411, 2016.

[134] A. Levant. Higher-order sliding modes, differentiation and output-feedback control. *Int. J. Control*, 76(9):924–941, 2003.

[135] Y. Shtessel, C. Edwards, L. Fridman, A. Levant. *Sliding Mode Control and Observation*. Springer, London, 2013.

[136] R. Su, Q. Zong, B.L. Tian, M. You. Comprehensive design of disturbance observer and non-singular terminal sliding mode control for reusable launch vehicles. *IET Control Theory Appl.*, 9(12):1821–1830, 2015.

[137] L. Wang, T.Y. Chai, L. Zhai. Neural-network-based terminal sliding-mode control of robotic manipulators including actuator dynamics. *IEEE Trans. Ind. Electron.*, 56(9):3296–3304, 2009.

[138] Y. Feng, X. Yu, Z. Man. Neural-network-based terminal sliding-mode control of robotic manipulators including actuator dynamics. *Automatica*, 38(12):2157–2167, 2002.

[139] Y. Feng, X. Yu, F. Han. On nonsingular terminal sliding-mode control of nonlinear systems. *Automatica*, 49(6):1715–1722, 2013.

[140] V.I. Utkin. *Sliding Modes in Control and Optimization*. Springer, Berlin, 1992.

[141] H. Alwi, C. Edwards. Fault detection and fault-tolerant control of a civil aircraft using a sliding-mode-based scheme. *IEEE Trans. Control Syst. Technol.*, 16(3):499–510, 2008.

[142] Q. Shen, D. Wang, S. Zhu, E.K. Poh. Inertia-free fault-tolerant spacecraft attitude tracking using control allocation. *Automatica*, 62(62):114–121, 2015.

[143] C.C. Coleman, F.A. Faruq. *On Stability and Control of Hypersonic Vehicles*. Defence Science and Technology Organization, Edinburgh, Australia, Tech. Rep. DSTO-TR-2358, 2009.

[144] J. Yang, S.H. Li, J.Y. Su, X.H. Yu. Continuous nonsingular terminal sliding mode control for systems with mismatched disturbances. *Automatica*, 49(7):2287–2291, 2013.

[145] B.L. Tian, L.H. Liu, H.C. Lu, Z.Y. Zuo. Multivariable finite time attitude control for quadrotor UAV: Theory and experimentation. *IEEE Trans. Ind. Electron.*, 65(3):2567–2577, 2018.

[146] F.Y. Chen, W. Lei, K. Zhang, G. Tao, B. Jiang. A novel nonlinear resilient control for a quadrotor UAV via backstepping control and nonlinear disturbance observer. *Nonlinear Dyn.*, 85(2):1281–1295, 2016.

[147] F.Y. Chen, R. Jiang, K. Zhang, B. Jiang, G. Tao. Robust backstepping sliding-mode control and observer-based fault estimation for a quadrotor UAV. *IEEE Trans. Ind. Electron.*, 63(8):5044–5056, 2016.

[148] F.Y. Chen, L. Cai, B. Jiang, G. Tao. Direct self-repairing control for a helicopter via quantum multi-model and disturbance observer. *Int. J. Syst. Sci.*, 47(3):533–543, 2016.

[149] F.Y. Chen, K.K. Zhang, B. Jiang, C.Y. Wen. Adaptive sliding mode observer-based robust fault reconstruction for a helicopter with actuator fault. *Asian J. Control*, 18(4):1558–1565, 2016.

[150] K.W. Lee, S.N. Singh. Robust higher-order sliding-mode finite-time control of aeroelastic systems. *AIAA J. Guid. Control Dyn.*, 37(5):1664–1671, 2014.

[151] J.L. Hunt, G. Laruelle, A. Wagner. *Systems challenges for hypersonic vehicles*. NASA Langley Research Center, Hampton, Virginia, Tech. Rep. NASA-TM-112908, 1997.

[152] E. Baumann, J.W. Pahle, M.C. Davis. The X-43A flush airdata sensing system flight test results. *AIAA J. Spacecr. Rockets*, 47(1):48–61, 2010.

[153] M. Basin, C.B. Panathula, Y. Shtessel. Multivariable continuous fixed-time second-order sliding mode control: Design and convergence time estimation. *IET Control Theory Appl.*, 11(8):1104–1111, 2017.

[154] B.L. Tian, Z.Y. Zuo, X.M. Yan, H. Wang. A fixed-time output feedback control scheme for double integrator systems. *Automatica*, 80(80):17–24, 2017.

[155] M.T. Angulo, J.A. Moreno, L. Fridman. Robust exact uniformly convergent arbitrary order differentiator. *Automatica*, 49(8):2489–2495, 2013.

[156] B. Fidan, M. Mirmirani, P.A. Ioannou. Flight dynamics and control of air-breathing hypersonic vehicles: Review and new directions. In *Proceedings of 12th AIAA International Space Planes and Hypersonic Systems and Technologies*, pages 2003–7081. Virginia, USA, AIAA, 2003.

[157] B. Yao, M. Tomizuka. Smooth robust adaptive sliding mode control of manipulators with guaranteed transient performance. *J. Dyn. Syst. Meas. Contr.*, 118(4):764–775, 1996.

[158] O. Iloputaife. Minimizing pilot-induced-oscillation susceptibility during C-17 development. In *Proceeding of the 22nd AIAA Atmospheric Flight Mechanics Conference*, pages 155–169. New Orleans, USA, 1997.

[159] T.A. Nelson, R.A. Landes. Boeing 777 development and APC assessment. In *Proceedings of the Society of Automotive Engineers (SAE) Control and Guidance Systems Conference*. Salt Lake City, USA, 1996.

[160] M.A. Dornheim. Report pinpoints factors leading to YF-22 crash. *Avia. Week Space Technol.*, 137(19):53–54, 1992.

[161] A. Brandstrom. *Coping with a Credibility Crisis: The Stockholm JAS Flighter Crash of 1993*. Swedish National Defense College, Stockholm, Sweden, 2001.

[162] P. Goupil. Oscillatory failure case detection in the A380 electrical flight control system by analytical redundancy. *Control Eng. Pract.*, 18(9):1110–1119, 2010.

[163] B. Lu, F. Wu, S.W. Kim. Linear parameter-varying anti-windup compensation for enhanced flight control performance. *AIAA J. Guid. Control Dyn.*, 28(3):494–505, 2005.

[164] E. Lavretsky, N. Hovakimyan. Stable adaptation in the presence of actuator constraints with flight control applications. *AIAA J. Guid. Control Dyn.*, 30(2):337–345, 2007.

[165] N.E. Kahveci, P.A. Ioannou, M.D. Mirmirani. Adaptive LQ control with anti-windup augmentation to optimize UAV performance in autonomous soaring applications. *IEEE Trans. Control Syst. Technol.*, 16(4):691–707, 2008.

[166] Z.T. Dydek, A.M. Annaswamy, E. Lavretsky. Adaptive control and the NASA X-15-3 flight revisited. *IEEE Control Syst. Mag.*, 30(3):32–48, 2010.

[167] C. Roos, J.M. Biannic, S. Tarbouriech, C. Prieur, M. Jeanneau. On-ground aircraft control design using a parameter-varying anti-windup approach. *Aerosp. Sci. Technol.*, 14(7):459–471, 2010.

[168] R.A. Hess, S.A. Snell. Flight control system design with rate saturating actuators. *AIAA J. Guid. Control Dyn.*, 20(1):90–96, 1997.

[169] S.A. Snell, R.A. Hess. Robust, decoupled, flight control design with rate saturating actuators. *AIAA J. Guid. Control Dyn.*, 21(3):361–367, 1998.

[170] Y. Zeyada, R.A. Hess, W. Siwakosit. Aircraft handling qualities and pilot-induced oscillation tendencies with actuator saturation. *AIAA J. Guid. Control Dyn.*, 22(6):852–861, 1999.

[171] J.M. Biannic, S. Tarbouriech. Optimization and implementation of dynamic anti-windup compensators with multiple saturations in flight control systems. *Control Eng. Pract.*, 17(6):703–713, 2009.

[172] X.G. Yan, C. Edwards. Robust sliding mode observer-based actuator fault detection and isolation for a class of nonlinear systems. *Int. J. Syst. Sci.*, 39(4):349–359, 2008.

[173] B.J. Bacon, A.J. Ostroff, S.M. Joshi. Reconfigurable NDI controller using inertial sensor failure detection & isolation. *IEEE Trans. Aerosp. Electron. Syst.*, 37(4):1373–1383, 2001.

[174] H. Alwi, C. Edwards. Robust sensor fault estimation for tolerant control of a civil aircraft using sliding modes. In *Proceedings of the American Control Conference (ACC)*, pages 5704–5709. Minneapolis, USA, 1996.

[175] B. Zhu, X. Wang, K.Y. Cai. Tracking control for angular-rate-sensorless vertical take-off and landing aircraft in the presence of angular-position measurement. *IET Control Theory Appl.*, 4(6):957–969, 2010.

[176] S. Hussain, M. Mokhtar, J.M. Howe. Sensor failure detection, identification, and accommodation using fully connected cascade neural network. *IEEE Trans. Ind. Electron.*, 62(3):1683–1692, 2015.

[177] A. Marcos, G.J. Balas. *A Boeing 747-100/200 aircraft fault tolerant and fault diagnostic benchmark.* University of Minnesota, Minnesota, USA, Tech. Rep. AEM-UoM-2003-1, 2003.

[178] H. Alwi,, C. Edwards. Fault tolerant longitudinal aircraft control using nonlinear integral sliding mode. *IET Control Theory Appl.*, 8(17):1803–1814, 2014.

[179] T. Wang, W.F. Xie, Y.M. Zhang. Sliding mode fault tolerant control dealing with modeling uncertainties and actuator faults. *ISA Trans.*, 51(3):386–392, 2012.

[180] C. Hanke, D. Nordwall. *The simulation of a jumbo jet transport aircraft, Volume II: Modelling data.* Boeing Commercial Airplane Company, Seattle, USA, Tech. Rep. CR-114494/D6-30643-VOL2, 1970.

[181] C. Edwards, T. Lombaerts, H. Smaili. *Fault Tolerant Flight Control: A Benchmark Challenge.* Springer-Verlag, Berlin, 2010.

[182] A.D. King. Inertial navigation – forty years of evolution. *GEC Rev.*, 13(3):140–149, 1998.

[183] US Dynamics. *US Dynamics model 475 series rate gyroscope technical brief.* US Dynamics Corp., New York, USA, Tech. Rep. USD-AN-001, 2005.

[184] C. Edwards, S.K. Spurgeon. *Sliding Mode Control: Theory and Applications.* CRC Press, London, 1998.

[185] V. Utkin, J. Guldner, J. Shi. *Sliding Mode Control in Electromechanical Systems.* Taylor and Francis, New York, 1999.

[186] V. Utkin. Variable structure systems with sliding modes. *IEEE Trans. Autom. Control*, 22:212–222, 1977.

[187] O. Harkegard, S.T. Glad. Resolving actuator redundancy – Optimal control vs. control allocation. *Automatica*, 41(1):137–144, 2005.

[188] L.J. Chen, H. Alwi, C. Edwards. Application and evaluation of an LPV integral sliding mode fault tolerant control scheme on the reconfigure benchmark. In *Proceedings of the American Control Conference (ACC)*, pages 3692–3697. Boston, USA, 2016.

[189] L.J. Chen, R. Patton, P. Goupil. Robust fault estimation using an LPV reference model: ADDSAFE benchmark case study. *Control Eng. Pract.*, 49:194–203, 2016.

[190] P. Goupil. AIRBUS state of the art and practices on FDI and FTC in flight control system. *Control Eng. Pract.*, 19(6):524–539, 2011.

[191] P. Goupil, A. Marcos. The European ADDSAFE project: Industrial and academic efforts towards advanced fault diagnosis. *Control Eng. Pract.*, 31:109–125, 2014.

[192] H. Berghuis, H. Nijmeijer. A passivity approach to controller-observer design for robots. *IEEE Trans. Robotic Autom.*, 9(6):740–754, 1993.

[193] M. Kerr, P. Goupil, J. Boada-Bauxell, P. Rosa, C. Recupero. Reonfigure FP7 project preliminary results and contributions. In *Proceedings of the 3rd International Conference on Control and Fault-Tolerant Syst (SysTol)*, pages 777–782. Barcelona, Spain, 2016.

[194] K. Kang M. Pachter, C.H. Houpis. Modeling and control of an electro-hydrostatic actuator. *Int. J. Robust Nonlinear Control*, 7(6):591–608, 1997.

9780367701796